SURFACE PROPERTIES OF SEMICONDUCTORS

POVERKHNOSTNYE SVOISTVA POLUPROVODNIKOV

ПОВЕРХНОСТНЫЕ СВОЙСТВА ПОЛУПРОВОДНИКОВ

SURFACE PROPERTIES
of
SEMICONDUCTORS

Edited by
Acad. A.N. FRUMKIN

Authorized translation from the Russian by
A. TYBULEWICZ, B.Sc., A.Inst.P., M.I.Inf.Sc., F.I.L.

Springer Science+Business Media, LLC 1964

ISBN 978-1-4899-5279-0   ISBN 978-1-4899-5277-6 (eBook)
DOI 10.1007/978-1-4899-5277-6

Library of Congress Catalog Card Number 63-21219
© 1964 Springer Science+Business Media New York
Originally published by Consultants Bureau Enterprises, Inc. in 1964
Softcover reprint of the hardcover 1st edition 1964

PREFACE

The present collective work is devoted mainly to studies of the effect of the surface state of semiconductors, principally germanium and silicon, on their electrical properties.

The volume includes contributions on the investigation, using various methods, of both clean surfaces and "real" surfaces covered with oxide films.

Studies of adsorption processes on clean surfaces and the influence of adsorbed oxygen on the electrical properties help to determine the mechanism of formation of oxide layers on the surfaces of germanium and silicon, as well as the mechanism of adsorption of various substances on semiconductor surfaces.

A considerable part of the volume deals with studies of the electrical properties of semiconductors governed by the existence of local electron states on the semiconductor surface.

Surface electron processes in semiconductors have been investigated in great detail in recent years and this is reflected in the present volume.

These processes are also of interest because they determine many properties of semiconducting devices. The poor reproducibility and stability of the surface charge and the surface recombination velocity make the parameters of these devices nonreproducible and unstable. However, the development of methods of surface treatment for semiconductors and semiconducting devices, which would ensure the desired stability and reproducibility of the surface properties, is possible only if the nature and laws governing surface electron properties are understood.

The volume consists of the papers read at the Conference on the Surface Properties of Semiconductors, held on June 5 and 6, 1961, at the Institute of Electrochemistry of the USSR Academy of Sciences, Moscow.

Academician A.N. Frumkin

PUBLISHER'S NOTE

The following Soviet journals cited in this book are available in cover-to-cover translations:

Russian title	English title	Publisher
Doklady Akademii Nauk SSSR	Proceedings of the Academy of Sciences of the USSR, Section: Chemical Technology	Consultants Bureau
	Proceedings of the Academy of Sciences of the USSR, Section: Chemistry	Consultants Bureau
	Proceedings of the Academy of Sciences of the USSR, Section: Physical Chemistry	Consultants Bureau
	Soviet Physics—Doklady	American Institute of Physics
Fizika tverdogo tela	Soviet Physics—Solid State	American Institute of Physics
Izvestiya Akademii Nauk SSSR: Otdelenie khimicheskikh nauk	Bulletin of the Academy of Sciences of the USSR: Division of Chemical Sciences	Consultants Bureau
Izvestiya Akademii Nauk SSSR: Seriya fizicheskaya	Bulletin of the Academy of Sciences of the USSR: Physical Series	Columbia Technical Translations
Kinetika i kataliz	Kinetics and Catalysis	Consultants Bureau
Optika i spektroskopiya	Optics and Spectroscopy	American Institute of Physics
Pribory i tekhnika éksperimenta	Instruments and Experimental Techniques	Instrument Society of America
Radiotekhnika i élektronika	Radio Engineering and Electronics	Massachusetts Institute of Technology
Zhurnal fizicheskoi khimii	Russian Journal of Physical Chemistry	The Chemical Society (London)
Zhurnal tekhnicheskoi fiziki	Soviet Physics—Technical Physics	American Institute of Physics

CONTENTS

I

EFFECT OF ADSORBED ATOMS AND MOLECULES ON THE SURFACE PROPERTIES OF SEMICONDUCTORS

DEFECTS AND IMPURITIES IN SEMICONDUCTORS
AND THEIR ROLE IN CHEMICAL TRANSFORMATIONS

S. Z. Roginskii

Chemical Physics Institute, Academy of Sciences, USSR

One of the by-products of the participation of chemists in the work on the preparation and practical applications of new semiconductors has been the infusion of new ideas and new methods into solid-state and macromolecular chemistry. A general survey of this influence on chemical kinetics was given by the present author at the end of 1960 [1]. These new ideas are directly related to the subject of our conference, since they represent a new approach to the kinetics of the formation and destruction of surface layers, to chemisorption, to the synthesis of semiconducting compounds, and to several other typical problems of semiconductor chemistry. In the present paper I shall deal with the role of crystal-structure defects and of microimpurities in chemical reactions in semiconductors. In accordance with the program of our conference I shall pay special attention to processes occurring or beginning at the surface.

1. Volume and Surface Types of Defects and Impurities

The fixed positions of the majority of molecules and atoms in solids and the short-range and long-range order make for a very great diversity of stable and "frozen-in" defects (related to previous history); these and departures from chemical stoichiometry and purity are present in practically every solid. Some of these imperfections are thermodynamically stable. Others are stable only in the dynamic sense, while still others are unstable [2].

Figure 1 shows a macroscopic defect related to previous history: a screw growth dislocation on the surface of SiC; the mechanism of its formation is shown in Fig. 2. Figure 3 shows schematically a homogeneous distribution of microimpurities (Li_2O and MgO) in solid solutions of donor and acceptor ions in NiO, which the author has investigated in some detail [3]. Figure 4 shows classical Frenkel and Schottky lattice microdefects. The examples of defects shown in Figs. 1-4 give some idea of the diversity of the various types of defect which occur in semiconductors.

Even to list and describe briefly all the currently known types of deviations from ideal structure and from ideal chemical composition, and to give briefly their properties, would require far too much space. The present communication is therefore limited to the development of a classification, the beginnings of which are illustrated in the figures. All types of intermediate defects are known, from the macroscopic defects shown in Figs. 1 and 2, through microscopic defects, right down to defects of atomic dimensions (Fig. 4). The classification of defects according to their dimensions is not clear-cut, because of the displacements and deformations near defects of atomic and microscopic dimensions, as shown in Fig. 5 [3]. In the limiting cases there is a clear division between chemical defects, examples of which are shown in Fig. 3, and structural "physical" defects,* which are shown in Figs. 1, 4, and 5. Chemical defects may be atoms, simple and complex ions, and

---

*This division, sharp in the ideal case, becomes less distinct in real solids.

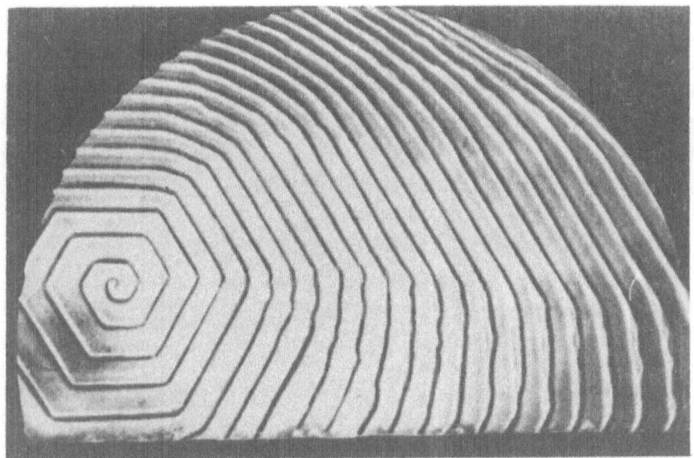

Fig. 1. Emergence of a screw growth dislocation on the surface of silicon carbide.

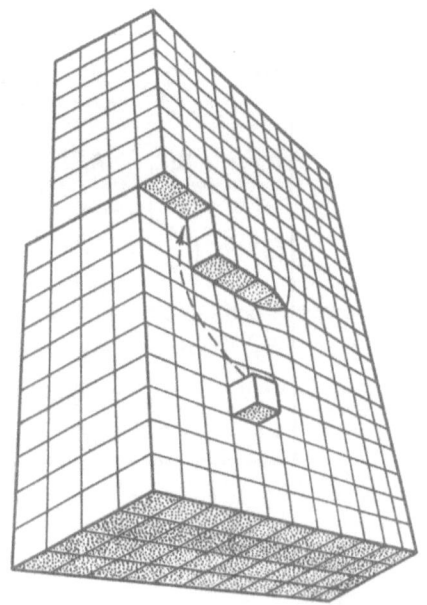

Fig. 2. Formation of a growth dislocation according to Frank.

Ni²⁺	O²⁻	Li⁺	O²⁻	Ni²⁺	O²⁻
O²⁻	Ni²⁺	O²⁻	Ni³⁺	O²⁻	Ni²⁺
Ni²⁺	O²⁻	Ni²⁺	O²⁻	Ni²⁺	O²⁻
O²⁻	Ni²⁺	O²⁻	Mg²⁺	O²⁻	Ni²⁺
Ni²⁺	O²⁻	Ni²⁺	O²⁻	Ni²⁺	O²⁻

Fig. 3. Chemical impurity substitution defects in nickel oxide.

molecules of foreign impurities (cf. Figs. 3 and 6: Sb and In in Ge), as well as constituents particles of the matrix lattice in abnormal valence states, for example Ni^{3+} in NiO or Zn^+ and Zn in ZnO.

A very frequent and important type of defect is the boundary between crystallites within a crystal and the external surface of such crystallites. For ferroelectric and ferromagnetic materials we can add here domain walls. In an idealized theory the surface of a solid is considered to be a geometrical discontinuity of the lattice accompanied by the appearance of additional Tamm [2, 4] and Shockley [5] surface levels. The number and positions of such levels can, in principle, be calculated if the volume system of energy levels of the crystal is known [6]. Such an ideal surface, which has obvious uses in the calculation of theoretical estimates, is a poor representation of the physical properties and special structural features of real surfaces. Apart from the ideal surface electron-energy levels, which are due to the discontinuity of the lattice, real surfaces frequently have coarser features of a strongly marked chemical nature. Examples of such features are oxide films on the surface of Ge and Si, shown in Fig. 7, which play a very important role in the surface properties of these semiconductors.

An ideally clean and perfect surface, free from microimpurities, defects, projections, recesses, and vacancies, i.e., an ideally flat lattice with strict periodicity, would be unstable under real conditions and would require special measures for its storage.

In recent years high-vacuum technique and new methods of surface cleaning and leveling have made great strides toward achieving an ideal state of the surface of metals and elemental semiconductors. As an illustration we give in Fig. 8 the field-emission photomicrograph of a single-crystal Si point, recorded by Allen[8].

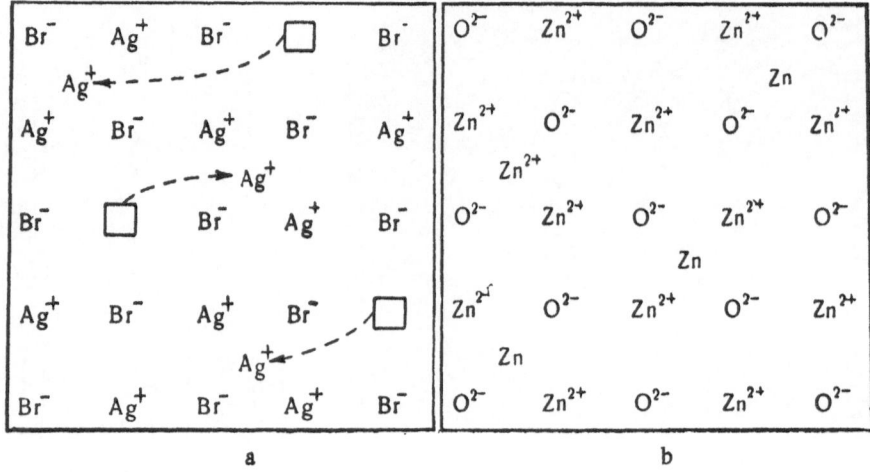

Fig. 4. Examples of crystal lattices with different types of defects: a) Frenkel defects in AgBr; b) Schottky defects in ZnO; c) ion vacancies of one sign in FeS; d) ion vacancies of two signs in NaCl

Because of the low melting point of semiconductors, which facilitates atomic creep and the transformation of the lattice on heating, it has not yet been possible to obtain ion photomicrographs of semiconductor points which show directly the positions of separate atoms in the surface layer. Such photomicrographs for metal points, obtained by several workers, have shown that even in annealed single crystals, with dimensions of the order of several hundred angstroms, there are departures from ideal periodicity of the atomic distribution [7]. This is also true of organic crystals, which have attracted the attention of Mentor and others working on semiconductors [8].

Departures from the ideal state increase enormously even after short exposure of metal and semiconductor points to the usual gases and vapors (O_2, N_2, H_2O) at pressures of the order of 10^{-6}-10^{-7} mm Hg. Stray projections of very small dimensions (of the order of tens of angstroms) are easily formed on the surfaces of points in strong electric fields.

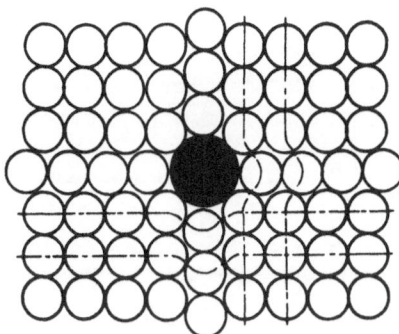

Fig. 5. Lattice distortions due to the presence of one larger particle.

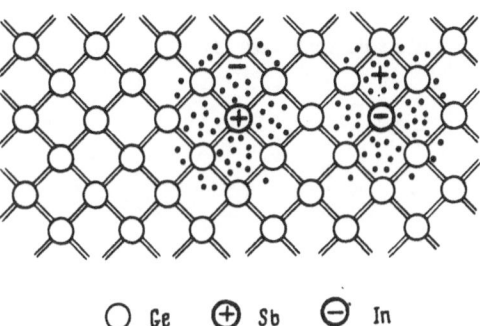

○ Ge   ⊕ Sb   ⊖ In

Fig. 6. Solid solutions in germanium.

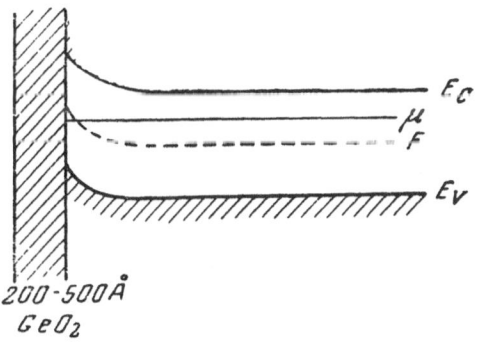

200-500Å
GeO₂

Fig. 7. Surface layers of germanium oxide on germanium.

Strong electric fields and heating are not the only external agencies which produce new types of defects. Similar effects are produced by illumination, irradiation with energetic electrons and heavy particles, etc. A characteristic example of such a defect is a thermal spike formed at the end of the trajectory of a high-energy ion in a solid. Such spikes have high concentrations of structural and chemical defects.

External and internal surfaces in a solid not only concentrate the defects, and establish conditions favorable for the appearance of special surface defects, but also play an important role in the generation and annihilation of such defects. Impurities from outside pass through the surface into the surface layer and into the interior of a crystal, and impurities already present in a solid leave through its surface on special treatments. The surface and the layer beneath it are the regions of the greatest and most rapid changes in the chemical composition. Deviations from the ideal order in the lattice, in particular single vacancies, diatomic and larger gaps, departures from the correct order of A and B atoms in AB-type lattices, etc., are all easily generated and annihilated at the surface. Finally the surface is the region where the electron and hole densities are different, and where there are thermodynamically stable deviations from the impurity concentrations in the interior of the crystal.

Impurities can be surface-active, accumulating in the surface layer, or surface-inactive, with depleted concentration in the surface layer. The volume and surface solubilities of impurities are, in general, unequal even in perfect crystals. As an example we shall quote the results obtained in our laboratory during studies of the influence of various impurities, dissolved in nickel oxide, on the electrical conductivity σ and the work function φ of this material. The former rises rapidly on introduction of monovalent metal oxides (for example, Li_2O), decreases sharply on introduction of trivalent metal oxides (for example, Ga_2O_3) and is practically unaffected by the introduction of MgO and ZnO (Fig. 9) [9]. In view of the p-type conduction of NiO, this represents a lowering of the Fermi level on introduction of Me_2O, its raising under the influence of Me_2O_3 and the retention of its position unaltered on introduction of MeO. Such shifts of the volume Fermi level should lead to an increase of the work function φ under the influence of Me_2O, its reduction due to Me_2O_3 and no effect on introduction of MeO. In fact the introduction of Li_2O reduced φ, while Me_2O_3 increased φ; MeO also increased it but very slightly (Fig. 10). Since the work function was determined by comparison with the same vibrating gold electrode [10], measurements of the changes of the contact potential difference on introduction of admixtures into pure nickel oxide gave reliable results. X-ray diffraction patterns confirmed the presence of NiO + MgO and NiO + Li_2O volume solid solutions in our samples. For such conditions, the relationship between

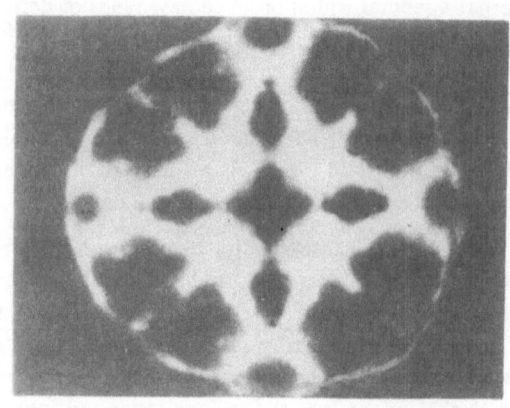

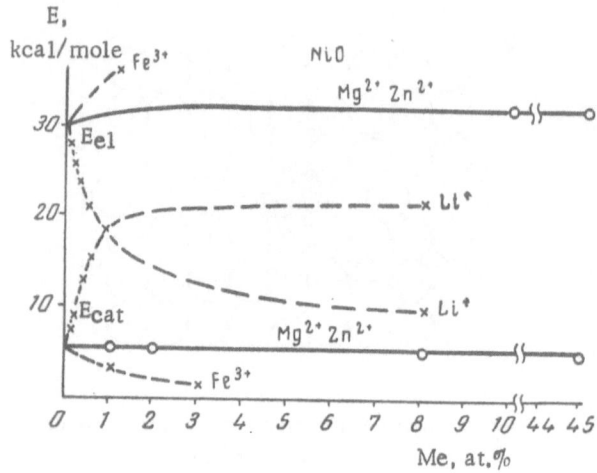

Fig. 8. Field-emission photomicrograph of a clean surface of a single-crystal Si point.

Fig. 9. Variation of the activation energy and of E_{cat} with the composition of pure nickel oxide and of its solid solutions with Li, Mg, Fe, and Zn.

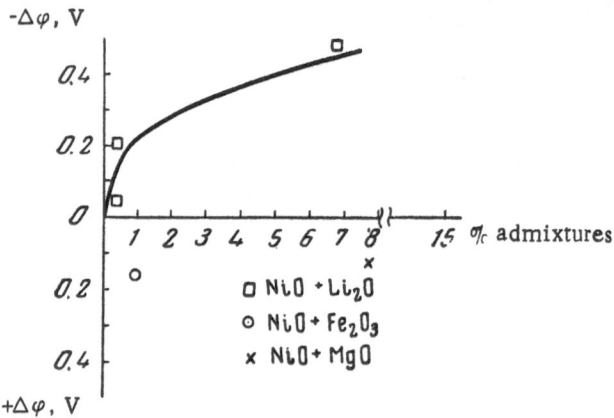

Fig. 10. Effect of Li, Mg, and Fe admixtures on the work function of NiO.

the surface and the volume electronic properties is more complex than that usually assumed in the attempts — as yet very inadequate — to develop an electronic theory of chemisorption and catalysis on semiconductors.

Similar results were obtained for the influence of these and other admixtures on the work function φ of zinc oxide. In this case we failed to detect solid solutions with Li_2O. The results obtained at our laboratory on powders agree with the results published recently for ZnO single crystals [11]. All these data on the influence of differently charged oxide admixtures in divalent metal oxides lead us to conclude that the effect of such admixtures on the surface electronic properties of semiconducting oxides cannot be reduced simply to a change in the carrier density in the interior corresponding to a change of the electrochemical surface potential.

As shown by Sh. M. Kogan and B. V. Sandomirskii in the development of their very useful phenomenological theory, some of the difficulties and disagreements with the usual theories can be avoided by assuming the existence of a wide spectrum of surface electron-energy levels, including latent levels which do not participate directly in chemisorption [12].

Very illuminating results, confirming the special nature of the relationships between the surface and volume properties of semiconductors, were obtained by V. M. Frolov in an investigation of the influence of acceptor and donor admixtures on the electronic and surface-chemical properties of germanium. Small admixtures of antimony and indium were introduced separately, in the form of suitable alloys, into pure molten germanium (electrical resistivity 10 $\Omega \cdot cm$ at room temperature). The electronic, catalytic and chemisorption properties of Ge thus doped were investigated by Frolov [13] after fragmentation of the crystal ingots and vacuum treatment. The Sb impurity enhanced the n-type conduction and In intensified the p-type conduction of Ge. The magnitude of the Fermi level shift on introduction of impurities (cf. Fig. 12) is well known from experiments on single crystals. Minute amounts of an impurity, for example 10^{15} cm^{-3}, i.e., 2×10^{-7} at. %, alter the catalytic activity of germanium considerably. Figure 11 shows this for the activation energy of ethyl alcohol dehydrogenation:

$$C_2H_5OH \rightarrow C_2H_4O + H_2 \qquad (1)$$

Even impurity concentrations of 10^{16} cm^{-3} were sufficient for the limiting values to be reached of the catalytic activation energies of n-type and p-type samples, which differed by 9 kcal/mole, i.e., by approximately half the forbidden band width of pure germanium. At the same time the Fermi level shift and the change in the volume electron density did not reach limiting values (cf. Fig. 12). As was first shown by Bardeen, a germanium surface having, in Frolov's case, less than one monolayer of oxygen before tests, is quasi-isolated in electronic properties [14]. This leads us to expect little or no influence of the volume Fermi level shift on the surface electronic processes. As has also been shown by other workers, this is true for the work function, which is altered by less than 0.01 eV on introduction of 10^{16} cm^{-3} of an impurity.

However, purely chemical properties, such as the catalytic activity and the related chemisorption of some gases, are very sensitive to volume admixtures. Thus in alcohol dehydrogenation, for example, which is a process of oxidation-reduction type involving electron transitions between the reacting molecules and semiconductors [15], the surface is not quasi-isolated from the volume in catalytic and chemisorption aspects, although it is quasi-isolated with respect to work function. This is not true of all reactions. In particular, hydrazine decomposition is not affected by acceptor and donor impurities [13]. Thus for NiO and ZnO the relationships between σ and φ are anomalous. However, there is an explicit and simple correlation between the impurity-induced changes of φ and E_{cat}, in spite of the complex relationship between σ and E_{cat} and the absence of quasi-isolation of the surfaces of these oxides from their volumes with respect to their electronic properties but not for some surface chemical reactions (alcohol dehydrogenation). It is possible that the germanium surface is quasi-isolated for other reactions, e.g., decomposition of hydrazine. The underlying causes of such relationships merit further study.

The concept of carrier traps is widely used in semiconductor physics. These defects differ from others in their ability to capture electrons or holes. Depending on the capture rate, we speak of slow and fast traps. The rates of capture of electrons and holes by traps may differ considerably from the rates of their liberation.

The existence of surface and volume traps has been proved for germanium, silicon and certain other semiconductors, and the difference between their effects on recombination and carrier mobility has been established. The defects which act as carrier traps are important in many physical and chemical processes in

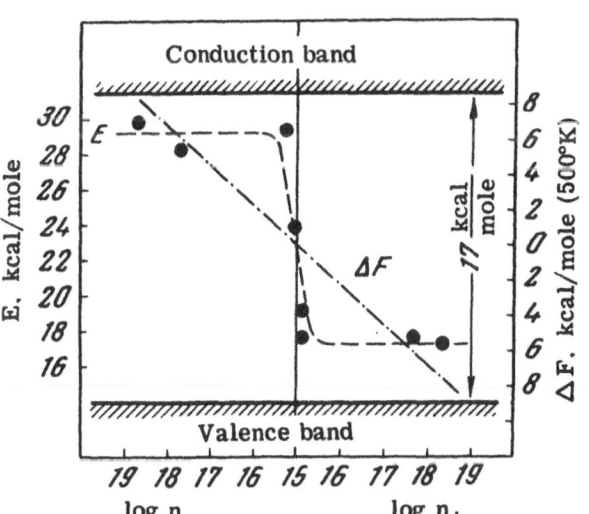

Fig. 11. Effect of indium (right) and antimony (left) admixtures in germanium on the position of the Fermi level (dashed) and the activation energy for the decomposition of ethanol (chain curve).

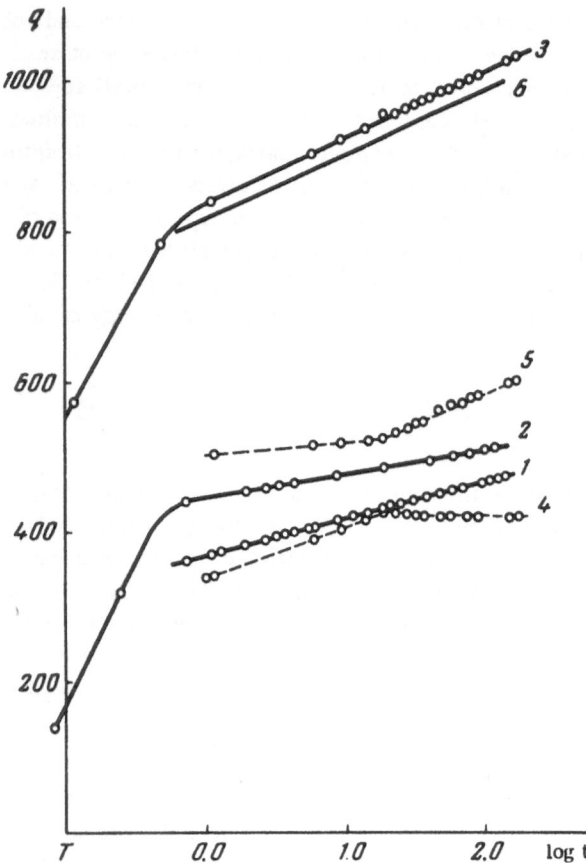

Fig. 12. Kinetics of adsorption of O_2, of pyridine, and of a (1 : 1) O_2 + pyridine mixture on NiO. 1) Adsorption of O_2 on vacuum-treated NiO, $p(O_2)$ = 0.54 mm Hg; 2) adsorption of pyridine on vacuum-treated NiO, p(pyridine) = 0.54 mm Hg; 3) adsorption of the (1 : 1) O_2 + pyridine mixture, total pressure 1.08 mm Hg, on vacuum-treated NiO; 4) adsorption of pyridine from the mixture; 5) adsorption of O_2 from the mixture; 6) calculated curve for additive adsorption of the mixture.

solids and in biological processes in large molecules. In particular their role in the initial stages of biological photosynthesis has been established [16]. Recently traps have been shown to have an important effect in catalysis [17]. Unfortunately the concept of a trap is frequently unrelated to specific defects. In the interior of solids and on their surfaces various types of defects can act as traps. Therefore, for a complete understanding of the role of traps and for control of the processes in which they participate, it is very important to find their physical and chemical nature in those cases where they are important. It is necessary to establish the properties of various defects acting as electron and hole traps in irradiated and unirradiated semiconductors. However, we meet with an inherent difficulty connected with the very frequent interrelationship between processes of generation of defects of various types. Thus, in particular, the absorption of a foreign atom or ion by a growing crystal is one of the typical causes of the appearance of growth dislocations. In turn, impurities are preferentially absorbed by dislocations of any origin and other structural defects form more easily there. A similar situation is true for vacancies in the lattice which are favored locations for impurity capture and for the appearance of particles with anomalous valence and anomalous charge.

A very important property of defects is their mobility. The most mobile are defects which differ only in their charge from the constituent particles or vacancies of the matrix lattice (electron — hole defects), and defects (excitons) which differ from the constituent particles and vacancies by occupying higher (excited) electron energy levels. The motion of electrons, holes, and excitons in the lattice, which is very important in normal as well as in photochemical and radiation-induced reactions, does not require the motion of heavy particles: atoms, ions, molecules. Therefore, the light particles move relatively quickly across the lattice from one site to another. This is helped by tunnel transitions characteristic of displacements of atomic order. Motion is also possible from trap to trap with intermediate transitions to the collective electrical conduction and excitation levels. The latter involve a certain loss of energy. When an electron or a hole moves in the interior, or on the surface, or between the interior and the surface, the corresponding chemical defect also changes its position. Thus the motion of a hole or an electron with temporary localization at individual particles in semiconductors with anomalously low carrier mobility (NiO, Fe_2O_3, etc.) represents the motion through the lattice of Ni^{3+} or Fe^{2+} without actual change of position of the corresponding nuclei, because in the lattices of NiO and Fe_2O_3 there are always Ni^{2+} and Fe^{3+} ions in the immediate vicinity of Ni^{3+} and Fe^{2+}.

The motion of a hole (+) or an electron (−) may take place by the tunnel effect over several atomic distances without considerable loss of the activation energy. The situation is different during the motion of a hole or an electron from trap to trap when there are few traps. Macroscopic processes caused by the motion of holes and electrons, for example changes in the surface charge, or the supply and removal of carriers to the reaction zone of some process, may be equally fast. Displacements of atoms, ions, and molecules are much more difficult and are slower. As first shown by Tamman and his school [18], a marked mobility of these heavy particles begins at temperatures when atoms in the lattice start changing places. In the absence of phase transition points these temperatures represent the beginning of firing conditions, amounting to some fraction of the absolute melting-point temperature: 0.33 for metals, 0.60-0.80 for other compounds. According to electron-microscopic data this is true also of the surface migration of particles [19]. A higher mobility also appears at phase transition points [18, 20]. The motion of the component particles of the lattice, appearing as self-diffusion, and the motion of foreign particles through dense lattices, are both directly related to the formation of vacancies* [21]. Usually the activation energies of the heavy-particle motion in the lattice are greater than those for light particles. In both cases the activation energies are usually smaller on the surface than in the interior.

For chemical processes taking place in the interior of a solid the multi-atom microscopic particles of microheterogeneous impurities, in the form of colloidal particles or microcrystallites of the foreign phase, can be regarded as fixed. On the surface, due to the planar creep involving "two-dimensional vapor" and due to slip, these particles can move both on the original matrix and on a foreign surface, as shown clearly by electron microscopy of deposits obtained by evaporation in vacuum [22].

2. Chemisorption

In all cases where the semiconductor surface is in contact with a gaseous medium (even if the latter is highly rarefied, $p \leq 10^{-7}$ mm Hg), and especially in contact with a liquid or a vapor, various adsorption processes occur on the surface. Of greatest interest is chemical adsorption, which leads to the formation of two-dimensional surface compounds. In the ideal case this type of adsorption ends after the formation of one, two, or more rarely three, layers of the product. For greater thicknesses of the product layer, for example in the case of the formation of an oxide layer 50 A or more in thickness during the etching of germanium, we are, strictly speaking, dealing not with chemical adsorption but with the formation of three-dimensional crystalline or amorphous layers† [23]. In this case we have the usual three-dimensional chemical reactions which begin at the surface but may spread to the whole grain or crystal. Apart from its intrinsic importance, because of the resultant changes in the physical properties of the surface and (to a lesser extent) of the interior, chemical adsorption is also of interest as one of the first stages of catalysis, electrochemical surface processes and volume reactions in solids.

Since the 1930's it has been known that the thermodynamic and kinetic adsorption isotherms for oxide semiconductors do not as a rule obey the equations derived from Langmuir's theory of adsorption, in which the surface is regarded as a perfect uniform "chessboard" with an exact number of "squares" per unit area of the surface, the only interaction of adsorbed molecules with one another being their competition for the limited available space. With these assumptions, for a surface coverage $\Theta < 0.05$ (i.e., less than 5% of a monolayer) the departures of the isotherms from linearity should be negligibly small. Experiments have shown that even in this region of surface coverage, in the case of static $\Theta = \Theta(c)$ and kinetic $\Theta = \Theta(t)$ isotherms, both Θ and the amount of the adsorbate q, which is proportional to it, rise more slowly than c or t, i.e., the isotherms are convex even for very small surface coverages. Also there is no tendency to saturation and the positions of the isotherms at Θ close to 1 are not horizontal, contradicting Langmuir's model [24]. By way of example we shall now quote the most typical chemical adsorption isotherms observed experimentally for semiconductors: static isotherms: $q \approx a \ln c$; $q \approx ac^{1/n}$; $\ln q = \ln a + (RT \ln c)^2$; kinetic isotherms: $q \approx a \ln t$; $q \approx at^{1/n}$; and others.

The observed departures from Langmuir's isotherms and the corresponding kinetics are due, in most cases, to a reduction of the enthalpy H of the true equilibrium or the enthalpy of the transient state on increase of the surface coverage. This is in accord with the reduction of the heat of adsorption (Q_a) and the increase of

---

*In zeolite structures ("molecular sieves"), which have not yet found direct application in the physics of semiconductors, the motion is possible along intrinsic permanent gaps and pores in the form of a characteristic chain of vacancies.

†Solution in a solid is also possible.

its activation energy (E_a) with increasing surface coverage [25]. At the same time Q_a decreases with increase of equilibrium values of c for static adsorption isotherms, and E_a increases with time in the case of adsorption kinetics. The dependence of the enthalpy on the surface coverage, $H = H(\Theta)$, may be due to previous surface inhomogeneity of H under equilibrium or transient conditions, and also to some mutual interaction between the adsorbed molecules (e.g., repulsion) which alters H. The simultaneous action of both these factors is also possible. The changes in the isotherms, as well as in Q_a and E_a, due to inhomogeneity cannot be distinguished from those due to repulsion. Therefore, the existence of surface inhomogeneity is not proved by experimentally obtained static and kinetic isotherms corresponding to varying H, nor by the establishment of varying isosteric H, nor even by the direct calorimetric proof of the reduction of the differential heats of adsorption with increase of the surface coverage.

Since 1945 we have developed in our laboratory several isotopic methods for detecting the inhomogeneity related to previous history both in its pure form and in conjunction with repulsion [26]. A particularly unambiguous and reliable method is the kinetic isotopic technique by means of which the presence of the inhomogeneity and the form of the surface distribution function of E_a is determined from the kinetic isotherms of isotopic exchange between adsorbed molecules and molecules in the gaseous phase.

A valuable supplement to isotopic methods of investigating surface inhomogeneity is provided by measurements of changes in the work function φ caused by adsorption. These measurements give us an estimate of the value of ΔH due to charging. In these measurements a serious difficulty is encountered in restoring the surface of a solid to a reproducible initial state. In particular, there are no certain methods of freeing the semiconductor surface from adsorbed substances and from substances dissolved in the surface layer. In binary compounds an attempt to do this by heating in high vacuum produces general or local changes in chemical composition. For example, ZnO decomposes partly into Zn, which remains in the lattice, and O_2 which is removed by pumping. In those cases (elemental semiconductors, chemically very stable compounds) where it is possible to remove the adsorbed substance from the surface by heating in ultra-high vacuum, the removal process unavoidably increases the number of structural defects. If bombardment with inert-gas ions is used, unstable solid solutions are formed as well. Therefore, with few exceptions, all the experimentally determined values of Θ are relative and not absolute. In the majority of methods we meet with complications due to changes in the potential of the second electrode which are typical of studies of reversible adsorption and have not yet been surmounted completely by anybody. When a semiconductor is in a gaseous medium all the methods of determining the absolute value of φ are unreliable. Only the change in the work function $\Delta\varphi$ can be found reliably and that only for the irreversible part of adsorption. The range of Θ available for studies can be increased by lowering T.

Thus very great experimental difficulties are met with in practical applications of isotopic and electronic methods to the study of surface inhomogeneities in chemical adsorption. In spite of this, some relationships between the inhomogeneity and the interaction (repulsion) between adsorbed molecules, which are important for the problems discussed here, have been established.

1.   As a rule when adsorption on a semiconductor is considerable, a large and sometimes dominant part of the chemical adsorption is irreversible or weakly reversible. This has been observed by É. Kh. Enikeev, L. N. Kurseva, Yu. N. Rufov, and others for several organic and inorganic gases and vapors adsorbed on some n- and p-type oxides and for some gases on germanium. Difficulties in completely freeing Ge and Si from chemisorbed oxygen and from dissolved CO and in obtaining clean surfaces are well known [27, 28]. In our laboratory we have investigated the effect of adsorption on surface charging in the systems listed below:

Semiconducting adsorbents: NiO; NiO with admixtures; ZnO; ZnO with admixtures; Ge; CuO; CuO with admixtures (NiO and CuO were studied in great detail).

Organic adsorbates: C_3H_6; C_2H_2; C_6H_6; pyridine; quinoline; alcohols; acetone.

Inorganic adsorbates: H_2; CO; O_2; N_2O; NO; CO_2; SO_2; N_2H_4; NH_3.

2.   All the isotopic and adsorption measurements indicated widespread surface inhomogeneity of the semiconductors studied in regard to the chemisorption activation energy, the isotopic-exchange activation energy, and the acceptor-donor properties.

3. The inhomogeneity appearing in these cases is first of all related to microchemistry; especially active are impurities and composition variations (differently charged ions in the lattice and on the surface, departures from stoichiometry, etc.) which alter the electronic properties of semiconductors.

4. Work on irradiated semiconductors (G. M. Zhabrova et al. [29]) has shown that this is also true for defects generated by irradiation, and particularly for secondary chemical effects of latent-image type, which are produced by oxidation-reduction processes in the lattice [30].

5. According to the published data, dislocations and vacancies are also active in Ge and Si, but their role has not been studied sufficiently fully for one to distinguish between the direct influence of such defects and the deformations due to impurities captured by these defects.

6. Apart from the effects of surface inhomogeneity, the kinetics of chemical adsorption is affected by self-charging of the surface by the sorbate, which in some cases (O_2 on ZnO) [31] accounts for a considerable part of the change in Q_a and E_a. This appears as a linear relationship between E_a, Q_a and the change in the work function $\Delta\varphi$ due to adsorption and the introduction of admixtures.

7. In the majority of cases particular molecules act in the same way on different surfaces. For example, O_2, N_2O, SO_2, and NO are electron acceptors, while benzene, acetone, pyridine, and quinoline are electron donors. A. N. Terenin and co-workers came to similar conclusions by determining $\Delta\varphi$ from the influence of adsorption on the external photoelectric effect [32].

Although all these adsorption results were obtained directly for individual molecular species, they may represent some latent and more complex processes, in particular the compensation of the charge of molecules already present on the surface, for example O_2; the adsorption with displacement of certain molecules, for example water; or chemical reactions with adsorbed molecules (in some cases also dissolving in the lattice and volume chemical reactions).

In heterogeneous catalysis the adsorption of mixtures can be in principle avoided only in simple reactions of homomolecular isotopic exchange. In practice during catalysis there are several (sometimes many) types of adsorbed molecules on the surface: substances present initially, intermediate products, contact poisons and activators, and final reaction products. Even in physical applications of semiconductors mixtures are adsorbed in certain stages of treatment, preparation, etching, and annealing, and frequently during the operation of semiconducting devices. Among such mixtures are $H_2O + O_2 + CO_2$ adsorbed from air, $CO + H_2O$, organic compounds $+ O_2$, etc.

In 1947 the present author noted an important gap in the literature on adsorption: no work had been done on the kinetics of adsorbed mixtures [24]. This gap has been filled only very recently. Little is known also of the equilibrium adsorption of chemisorbed mixtures [33]. Recently this gap was partly filled by work carried out in our laboratory: Yu. N. Rufov investigated the chemisorption of various adsorbate pairs: donor and donor; acceptor and acceptor; acceptor and donor. The adsorption took place on oxide semiconductors. In the majority of cases a considerable part of the adsorption occurred very rapidly and this caused a fundamental difficulty in the study of the kinetics. On the other hand there were no difficulties in investigations of the subsequent slower stages of the adsorption process. For the slow processes, to a first approximation, the rates of adsorption of mixtures differed little from the rate of adsorption of the component which on its own is adsorbed faster; the rate depends very weakly on the partial and total pressures, and there is no clear difference between pairs of different electronic type, etc. These and other new relationships (some of them unexpected) obviously cannot be explained without hypotheses on the mechanism of chemisorption processes. Before considering these hypotheses we must stress that they are only working hypotheses to be rejected or considerably modified after more systematic investigations. We shall consider two approaches to the results obtained:

1) we may assume that the whole kinetic isotherm from several seconds to tens of hours (which was the maximum duration of our tests) represents one process;

2) the observed chemisorption represents two (or more) processes of different nature and occurring at different times.

The greatest interest was in systems in which, according to Rufov's experiments, the rate of chemisorption of the binary mixture was higher than the sum of the rates of the two components on their own. In one case (Fig. 12) one of the components of the mixture was an acceptor (O_2) and the other was a donor (pyridine), and therefore one could assume mutual interaction because the inhibition caused by charging was thus removed [31]. However, in another case (Fig. 13) both O_2 and SO_2 are acceptors on their own, and this simple explanation cannot be used. In the majority of cases the adsorption of mixtures consisting either of components of the same type (both donors or both acceptors) or of different types (donors and acceptors) proceeded at a rate much smaller than the sum of the rates of the components on their own. Even very small amounts of the gas or vapor chemisorbed during the first seconds produced large changes of φ. Such kinetic observations are not compatible with the representations of the chemisorption mechanism which follow from the boundary-layer theory of Schottky, Aigrain, Hauffe, and Garrett. In fact, in this theory a transition of one electron (or hole) to a lower level and the consequent charging of the surface are regarded as the main reason for chemisorption; the steady-state formation of a charged transient complex should therefore be sensitive to the sign and magnitude of $\Delta\varphi$ due to adsorption of the second component. Without additional assumptions the observed behavior cannot be made to agree with the less rigid representations of Vol'kenshtein, Kogan, and Sandomirskii. The latter workers assumed that the same molecules can have different types of bonding to the surface, some of which do not shift the surface Fermi level, and that molecules chemisorbed in different ways are in dynamic equilibrium with one another.

With this assumption as a starting point, important corrections are needed to another (very popular) representation of adsorbed molecules as surface defects which influence the system of electron levels and can act as new impurity centers. This influence may be considerably weaker than the influence of impurity centers produced by nonvolatile modifiers with differently charged particles because the latter, according to the available observations, when present in the same amounts and producing the same changes of φ, may have a considerably greater effect on the rate of chemisorption. * Some difficulties are avoided by the second of the two theories mentioned above, if, for example, we assume that on the surface of semiconductors subjected to heating in high vacuum we have at first a fast, practically irreversible, process which produces a surface coverage representing a considerable fraction of a monolayer. Such a process regenerates the initial surface destroyed by the vacuum treatment or forms a new surface on which the slower

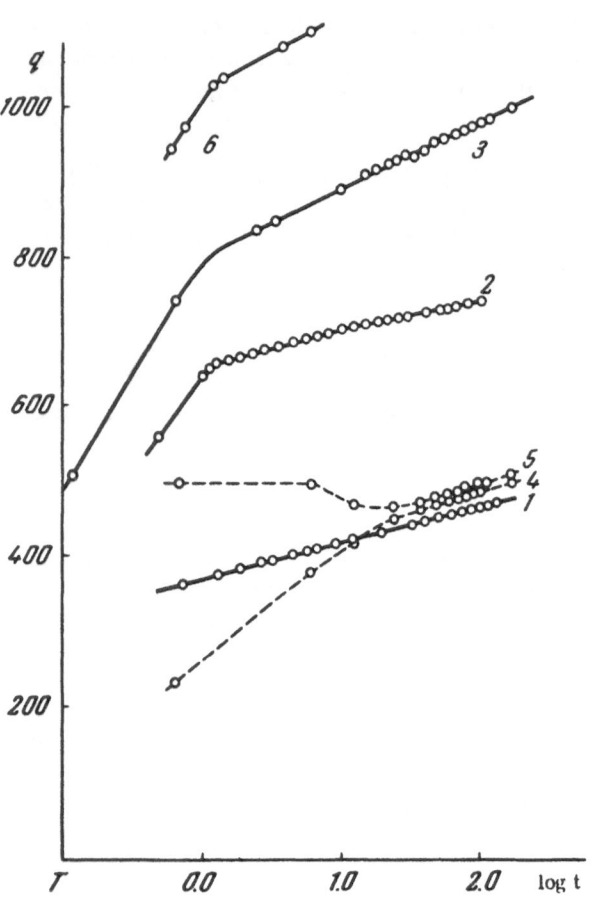

Fig. 13. Kinetics of adsorption of O_2 and SO_2 separately and as a (1 : 1) mixture on NiO. 1) O_2 on vacuum-treated NiO, $p(O_2)$ = 0.54 mm Hg; 2) SO_2 on vacuum-treated NiO, $p(SO_2)$ = 0.55 mm Hg; 3) $O_2 + SO_2$ (1 : 1) mixture, total pressure 1.1 mm Hg; 4) adsorption of SO_2 from the mixture; 5) adsorption of O_2 from the mixture; 6) calculated curve for additive adsorption of the mixture.

---

*The effect on chemisorption is frequently exaggerated. For example, Chizhikova and Keier [34] reported the complete cessation of chemisorption of CO on introduction of a small amount of Li_2O into ZnO. In later work we could not reproduce this result and it is not compatible with the idea that the chemisorption of CO controls the latter's oxidation to CO_2.

chemisorption occurs, and this process should not be taken into account directly in an analysis of the kinetic isotherms. On this assumption we should compare principally the parts of the isotherms representing the slow chemisorption, i.e., not q(t) but $q(t)_{slow} = q(t) - q(\tau)$. The weak point of this second hypothesis is the indeterminacy of the boundary separating the two parts of the isotherms.

The mutual acceleration of adsorption is met with frequently and is more considerable in the second theory, but so far we have not been able to establish a clear relationship between the similarity or dissimilarity of acceptor-donor properties of the two components of the binary mixture and the superadditive mutual acceleration.

In both theoretical approaches the experimental data on the kinetics of chemisorption of mixtures provide additional confirmation of the importance of defects (especially chemical ones) in the kinetics of chemisorption. This, and not the long-range effects, accounts for the strong poisoning effect of small amounts ($\Theta < 0.1$ monolayer) of chemisorbed CO on the rate of chemisorption of pyridine.

At various times the characteristic features of the kinetics of chemisorption on active surfaces have been explained by:

a) widespread surface inhomogeneity in regard to E_a and Q_a, due to previous history, and various functional relationships between these two energies [24];

b) surfaces with ΔH varying with surface coverage [24-26];

c) the influence of surface charging of semiconductors and metals on E_a and Q_a, which leads to expressions of the following type [35]:

$$E_a = E_0 \pm \alpha \, \Delta\varphi \text{ and } Q_a = Q_0 \pm \beta \, \Delta\varphi. \tag{2}$$

The hypotheses given under (a) - (c) have been confirmed by experiments. However, the kinetics of the chemisorption of mixtures cannot be explained completely by means of these ideas.

In contrast to the usually accepted view, we shall assume that

1) activation in the stage which controls the rate of chemisorption is related to the solid and represents the formation of some special local structures or local states C^*;

2) chemisorption occurs only on contact of a gas molecule with such an activated region of the surface (directly on impact or after preliminary physical adsorption), cf. equations A and B below;

3) the surface is inhomogeneous and the chemisorption activation energies E_a are different for different regions:

$$A. \begin{cases} 1. & C_i^{\circ} + ex \to C_i^{\circ *} \,; \quad w_{i1} = k_{0i}\,[C_i^{\circ}]\,[ex]\,\exp\,[-\,E_i/RT]; \\ 2. & C_i^{\circ *} + A \to C_i^{*}A_{chem}; \quad w_{i2} = k_{i2}\,[C_i^{\circ *}]\,[A]; \\ 3. & C_i^{*} \to C_i^{\circ} + ex; \quad w_{i3} = k_{i3}\,[C_i^{\circ *}]. \end{cases}$$

$$B. \begin{cases} 1. & C_i^{\circ} + \Theta \,(or + \oplus) \to C_i^{\pm *}\,; \quad w_{i1} = k_{0i}[C_i^{\circ}]\,[\Theta]\,\exp\,[-\,E_i/RT]; \\ 2. & C_i^{\pm *} + A \to C_i A_{chem}^{\pm}; \quad w_{i2} = k_{i2}[C_i^{\pm *}]\,[A]; \\ 3. & C_i^{\pm *} \to C_i + \Theta \,(or + \oplus); \quad w_{i3} = k_{i3}[C_i^{\pm *}]. \end{cases}$$

The above equations show two of the possible variants. In both variants chemisorption occurs on impact but in the first case the activated state is neutral and in the second it is charged. C_i° represents a free adsorption center on the surface; A is the adsorbed molecule; Θ, $\oplus$, and ex represent, respectively, an electron, a hole, and an exciton. $C_i^{\circ *}$ is an activated neutral adsorption center formed with an energy loss E_i. $C_i^{\pm *}$ is a similar center but carrying a positive or a negative charge. When $w_{i3} \ll w_{i2}$ the rate of chemisorption is independent of pressure and equal to the rate w_{i1} when the reverse reactions are neglected. For the surface as a whole $W = \Sigma w_i = \Sigma w_{i1}$. When the function which gives the distribution of the regions C according to their E or k_1 is known, the above sum can be replaced by the integral $\int \rho(k_1)\,dk_1$ or $\int \rho(E_1)\,dE_1$.

From the first and second assumptions it follows that when $P \geq P_{min}$ there should be little or no pressure dependence of the chemisorption rate. This is frequently found for chemisorption occurring at not too low pressures.

The third assumption on the presence of a distribution of surface regions with different E_a gives a mechanism typical of chemisorption of surfaces with E varying with surface coverage. In the case where the initial stage of activation of the surface regions controls the rate of adsorption and is not accompanied by a redistribution of charge between the volume and surface of a crystal, charging of the surface should not affect the chemisorption rate. If such a redistribution of charge (electron transitions) does occur, surface charging will influence the chemisorption rate and the effects of surface inhomogeneity and of charging may be additive [31]. An example of activation without charging is the formation of a fixed or wandering exciton and the formation of Schottky and Frenkel defects. Activation with charging can occur when any such defects are generated and electron and hole transitions take place between the surface and the volume, for example, during the formation of a surface ion of different valence according to

$$Me^{n+} + \ominus \to Me^{(n-1)+} \text{ or } X^{2-} + \oplus \to X^-$$

or charging of a surface anion ("an") or cation ("cat") vacancy $\square$:

$$\square_{an} + \ominus \to \square_{an}^- ; \square_{cat} + \oplus \to \square_{cat}^+$$

Under these assumptions practically every active center appearing on the surface reacts, and the activation stage is the one which controls the total rate. However, at low pressures or in the case of an additional activation energy of formation, from an activated region and a chemisorbed molecule, of an "intermediate complex" or an "intermediate state" (denoted by IC), even the i-th group of regions should exhibit a dependence of the chemisorption rate on the pressure of the chemisorbed gas. The rate may be directly proportional to the first power, square root, etc., of the pressure. The first part of the chemisorption isotherm, which is difficult to resolve, may represent the action of regions which require practically no activation energy E_a. This type of activation mechanism should appear most clearly in the differences between the chemisorption rates of isotopic molecules singly and of isotopic mixtures, in the absence of the kinetic isotopic effect. This difference was found for H_2 by Parravano, Friederick, and Boudart [36].

When surface regions are activated without charging (e.g., A above) the acceptor and donor molecules may be adsorbed by the same centers and the differences between their rates of sorption should be small. In the simplest case the rate of adsorption of mixtures should differ little from the adsorption rate of the components by themselves at a pressure equal to the total pressure of the mixture. The reported influence of small amounts of CO should then be the result of modification of the surface by chemical reaction. Obviously we cannot avoid such a hypothesis or a slightly different hypothesis of special regions which act as "chemisorption gates" through which all molecules have to pass before they can spread throughout the surface. The problem of the real existence of such mechanisms and the extent to which they occur in the kinetics of chemisorption of mixtures and individual substances requires further study.

3. Heterogeneous Chemical Reactions

As pointed out at the beginning of the present paper, chemisorption is a necessary stage of several types of chemical reaction occurring on the surface and in the interior of a solid. We shall give now the briefest possible review of the role of defects in some of these reactions.

1. Heterogeneous Catalysis

Certain forms of chemisorption are necessary stages in contact processes. Frequently the catalysis rate is controlled by the chemisorption of one of the components or the desorption of one of the chemisorbed products. Chemisorbed intermediate complexes, including IC during the controlling stages, are formed during catalysis.

With these relationships between catalysis and chemisorption the hypothesis of the control of the chemisorption rate by the formation of surface defects or the activation of surface centers should be extended to catalysis. In a study of the kinetic isotopic effects in the reaction $2H_2 + O_2 \rightarrow 2H_2O$ information has been recently obtained that such a mechanism is possible on platinum [37]. This observation and the similarity of the mechanism of catalysis on metals and semiconductors suggest that we should seek similar relationships in catalysis on semiconductors. In the 1880's D.P. Konovalov suggested experimental work to establish a relationship between the activity of solids and the excess energy of frozen-in active states. In the 1920's Taylor and others proposed a theory of active centers which has been generally accepted. In 1928 A. F. Ioffe indicated the possibility of a relationship between centers of catalytic activity and defects of crystal structure which affect the electrical conductivity [38]. In 1940 the present author proposed a microchemical concept of an active surface on which microimpurities and departures from stoichiometry would be particularly efficient in the formation of active centers [39]. Results confirming this concept for catalysis on semiconductors are given in Sec. 1 of the present paper; here we shall merely point out new facts which confirm and give a sounder basis to the idea of a relationship between catalytic activity and structure defects and microimpurities.

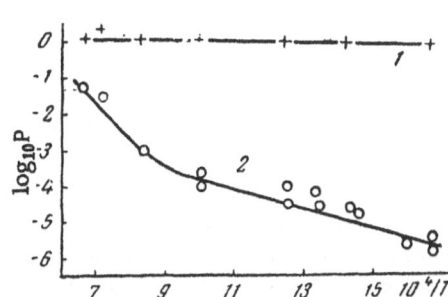

Fig. 14. Appearance, after pulse heating of catalytic "superactivity" for the decomposition of formic acid on metallic nickel: 1) superactive nickel; 2) active nickel.

Of great fundamental interest are the results of recent work on the "superactivity" of metal wires quenched after heating near the melting point [40]. By way of illustration we reproduce a figure (Fig. 14) from the latest work of Duell and Robertson [41]. In coordinates of $\log_{10}P$ and $10^4/T$ the figure gives the temperature dependence of the normal catalytic activity (the curve with a kink) and the superactivity of Ni (the upper horizontal line, $E = 0$); at low temperatures the absolute values of k of quenched Ni are 10^5-10^6 times greater than the activity of "normal" Ni. The enormous effect observed is related to the high defect concentration. Duell and Robertson considered that the defects are lattice vacancies which have emerged close to the surface. This conclusion does not seem to be entirely sound. Judging by the experimental conditions the microchemistry of the surface could have changed due to desorption and diffusion from the interior. Vacancies could have captured impurities from the gas and the metal could have been covered by a very thin semiconducting layer. Thus the nature of the defects is in doubt, though the relationship of the observed enormous effect to the presence of defects is certain. Single observations in our laboratory indicate the possibility of similar though less pronounced effects in germanium.

Valuable results, confirming the role of defects, have been obtained in the study of the effect of a preliminary irradiation of a solid (semiconductor or insulator) with fast electrons, gamma-rays or neutrons, on its catalytic activity. In several cases (SiO_2, ZnO, etc.) a considerable change in the catalytic activity was found to be due to relatively stable defects remaining after irradiation. In some cases the final activity was greater than the activity obtained by conditioning and nonirradiation methods [42]. Acceleration of the catalytic activity by irradiation of a solid during catalysis of gas reactions has also been observed [43]. This effect is also mainly due to a combined effect of the stable and unstable defects, the latter disappearing rapidly after the end of irradiation.

Recently several workers have questioned the idea of the role of active centers, and have asserted that the specific catalytic activities of catalysts are constant and independent of the method of preparation and subsequent treatment [44].

The results given above and many other reports disprove this point of view decisively, although in some cases contact with the reacting substances may strongly reduce or even destroy completely the initial differences between the catalytic activities.

The data available so far are insufficient to estimate the role and efficiency of various defects in radiation-chemical effects on catalytic activity. From the numerous available reports on the changes produced by the irradiation of solids it seems very likely that the stable radiation defects are chemical or structural defects locked by capture of impurities (see below). Effects of this type also occur during the so-called catalytic corrosion, i.e., during changes of the structure of a solid surface with appearance of faces with new indexes under the influence of catalytic reactions [45]. This may disrupt and disintegrate massive solids. For specialists in semiconductor science this phenomenon, the reverse of the influence of defects on the catalytic reaction, may be of definite interest. In operations such as the heating of semiconductors in gaseous media, etching of their surfaces with solutions containing corrosive reagents such as H_2O_2, etc., catalytic reactions are possible. Recent work has shown that many of the semiconductors which are of interest in technical applications (Ge, Si, and their isoelectronic analogs of $A^{III}B^V$ and $A^{II}B^{VI}$ type) and many spinels are active catalyzers of oxidation-reduction reactions [31]. Similar and stronger effects can probably be caused by chemical adsorption and desorption of oxygen on other substances. On the other hand the chemisorption is unavoidable, and it is not possible to exclude completely catalytic reactions occurring during the preparation and purification of semiconductors. Catalytic reactions are not only corrosive but are also capable of generating new defects on solid surfaces.

2. Chemical Reactions in Solids

Isolated observations indicating that defects are important in the kinetics of chemical reactions in solids were reported some time ago. Among these observations are the measurements of the induction period for dehydration of single crystals of some hydrates after mechanical damage to the surface. The reaction spreads from the damaged region (a scratch) throughout the surface and into the interior of the crystal. The localization of reactions at occlusions and defects is well known in corrosion. Mechanical deformation strongly affects the reactivity of inorganic and organic materials, and chemical damage is frequently concentrated at deformed locations. The magnitude of the calorimetrically measured excess energy of a solid has been found to be related to its reactivity [46]. During chemical reactions in solids anomalous physical properties have appeared which, according to current theories, indicate the generation of large numbers of defects [47]. For many reactions between solids and between solids and gases the controlling stage is the diffusion of certain ions and atoms in a foreign lattice, and all diffusion in crystals proceeds along existing defects or produces such defects at each elementary displacement step. Vacancies, formed by thermal motion, are primarily active in such diffusion. Frequently this diffusion is accelerated by the local electric fields [48].

For some decades the theory of chemical reaction kinetics has been dominated by the crystallization-chemical concept, according to which the kinetics is determined by formation of a new phase in the original solid and the growth of nuclei of this phase [49]. Some of the phenomena, ascribed on this concept to the autocatalytic action of nuclei of the new phase, are really due to defects generated by local stresses caused by the difference between the volumes, $\Delta V = V_{orig} - V_{fin}$, of the original and final phases. This difference may be positive or negative and it may sometimes reach several tens per cent. Under the influence of a nonzero value of ΔV, crystals may become fragmented, porous products may appear and internal stresses may be set up. In the case of semiconductors phenomena which appear to be crystallochemical may be due to the electron-modifying action of the contact between the original and new phases.

Reactions in solids take place in many stages. The accelerating effect of the nuclei of the new phase is a late macroscopic stage, which is preceded by microscopic stages [50]. This has been followed carefully for the decomposition of some solid salts accompanied by the evolution of gases. As an example of this we can take the decomposition of azides and oxalates of divalent metals:

$$Ba^{2+}(N = N = N)_2^- \to Ba + 3N_2;$$
$$Ni^{2+}(C_2O_4)^{2-} \to Ni + 2CO_2.$$

In the normal reactions the transformations of the unstable anion and the stable cation are to a large extent independent. As a rule these transformations are separated in space. In particular, during photochemical decomposition of azides we have first the formation of an exciton ("ex") or a hole ("p") and a conduction electron ("n"), depending on the energy of the incident quantum:

$$N_3^- + h\nu_1 \to N_3^{-*}, \quad \text{i.e.,} \quad N_3^- \ldots ex;$$
$$N_3^- + h\nu_2 \to N_3 + n, \quad \text{i.e.,} \quad N_3^- \ldots p + n, \text{where } \nu_2 > \nu_1.$$

Both an exciton and a hole are capable of moving through the crystal from one N_3^- anion to another and reaching the surface. Formally, this is equivalent to the motion of N_3 or N_3^{-*} to the surface (represented below by S–S). In fact all anions retain their position in the lattice:

$$N_3 \overset{p}{\to} N_3^- \overset{p}{\to} N_3^- \overset{p}{\to} N_3 \overset{S}{\underset{S}{\Big|}} = N_3^- N_3^- N_3^- N_3 \overset{S}{\underset{S}{\Big|}} ;$$

$$N_3^{-*} \overset{ex}{\to} N_3^- \overset{ex}{\to} N_3^- \overset{ex}{\to} N_3^- \overset{S}{\underset{S}{\Big|}} = N_3^- N_3^- N_3^- N_3^{-*} \overset{S}{\underset{S}{\Big|}} .$$

From the kinetic results it follows that the next stage for the surface decomposition of some azide ions and radicals occurs at sites of anion surface vacancies. In the case of exciton excitation, two excitons participate in the process, and due to the formation of the volatile product N_2 on the surface the number of vacancies increases [1, 51]:

$$\square_{an} + 2N_3^- + 2ex \to {}^*[N_3^- \square^- N_3]^* \to 3N_2 + 2\square_{an}^- + \square_{an}.$$

The reduction of metallic ions to atoms occurs mainly at anion vacancies and other defects which act as electron traps. The process occurs particularly easily on the surface at metallic nuclei because of the gain of energy on attachment of Me to the crystal:

$$\square_{an} + n \to \square_{an}^-; \quad 2\square_{an}^- + Me^{2+} \to Me.$$

The thermal decomposition of solids passes through similar stages. In all these cases there are characteristic effects of action at a distance. Difficulties of the emergence of the constituent particles of the lattice are overcome due to the relatively high mobility of excitons, holes, and electrons. In recent years the idea of the active participation of defects in chemical reactions in solids has gained more and more support [52]. Of decisive importance were the results obtained in investigations of the influence of corpuscular and hard photon (γ ray, x ray) radiation on chemical transformations in solids, since in these investigations it was possible to distinguish the stage of defect formation from the stages where these defects revealed their existence. We cannot deal in great detail with this problem and we shall mention only some investigations in which the mechanism of the participation of lattice defects has been proved with high reliability:

a) phase transition on irradiation of normal metallic tin into semiconducting grey tin [53];

b) oxidation of graphite with oxygen after large doses of neutron irradiation [54];

c) accelerated decomposition of copper and nickel oxalates after electron and neutron bombardment [29];

d) increase of the velocity of isotopic exchange $H_2 + D_2 \rightleftarrows 2HD$ and hydrogenation of ethylene on ZnO after irradiation [55].

The magnitude of these effects is shown in Fig. 15. As a rule the defects produced by irradiation increase both the electrical conductivity and the reactivity. In organic systems a characteristic large change in the magnetic properties of the irradiated material is observed, on polymerization in the solid phase, and in biochemical processes in biological polymers. In particular, characteristic electron paramagnetic resonance signals appear, indicating a relationship between the increased activity and the breaking of covalent bonds in molecules and the formation of radicals with high contents of unpaired electrons. Similar defects, although not in such high concentrations, can be formed in such organic systems without irradiation (under the influence of mechanical stresses, active impurities, etc.).

Very convincing results on the role of structural defects in polymerization processes have been obtained in particular by Academician V. A. Kargin, V. A. Kabanov, and N. A. Platé, and others [56]. Mechanical fragmentation of some frozen monomers, which generates defects detected by e.p.r. and coloring, makes these monomers capable of polymerization in the solid phase. These effects are even more pronounced on mechanical fragmentation of heterogeneous mixtures consisting of frozen monomers and alkali-halide crystals (NaCl, KCl and others) or semiconductor crystals (Ge, Si and others). Explosive polymerization at phase transition points is also related to the appearance at these points of a large number of vacancies together with an increased mobility of the molecules [57, 58].

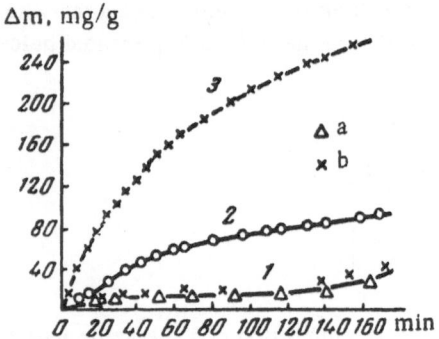

Fig. 15. Effect of preliminary irradiation on the kinetics of decomposition of copper oxalate in vacuum at 225°C.
1) Sample not subjected to irradiation (a) and sample which had absorbed a dose of 2.2×10^8 rad (b); 2) sample subjected to preliminary irradiation (6×10^8 rad); 3) sample subjected to preliminary irradiation (1.1×10^9 rad). The dose rate was $10 - 10^4$ rad/sec.

The importance of chemical impurities and structure defects in solids and their role as active centers in reactions are well established. However, in the majority of cases it is not clear which defects are active. Studies of the relative role of various defects in various types of processes and in special cases meet with experimental difficulties and are still in their infancy. The situation is similar in the problem of the importance of defects in the electronic physical properties of solids. In this problem the interests of chemists and solid-state physicists are closely interwoven, making this one more boundary between physics and chemistry which shows great promise for the control of physical and chemical processes in semiconductors. However, we must remember that in a solid the influence of a defect is related in a special way to cooperative defects due to the periodicity of the solid structure, and with macroscopic charging effects. We have no space to deal with this aspect, but without allowing for the collective and macroscopic effects in the action of microdefects we cannot obtain a correct picture of the role of defects in chemical reactions in solids [1, 58].

LITERATURE CITED

1. S. Z. Roginskii, Zhur. Vsesoyuz. Khim. Obshchestva im. D. I. Mendeleeva 5, No. 5, 482 (1960).
2. W. E. Garner, ed., Chemistry of Solid State [Russian Translation] (IL, 1961); H. G. Van Bueren, Imperfections in Crystals (Amsterdam, 1960); K. Hauffe, Reactions in and on Solids [Russian Translation] (IL, 1962).
3. S. Z. Roginskii, Khim. Nauka i Prom. 2, No. 2, 138 (1957).
4. I. Tamm, Phys. Z. Sowjetunion 1, 733 (1932).
5. W. Shockley, Phys. Rev. 56, 317 (1939).
6. Ya. Koutetskii, Kinetika i Kataliz 2, 319 (1961).
7. E. W. Müller, Advances in Electronics and Electron Phys. 13, 83 (1960).
8. J. W. Mentor, Proc. Roy. Soc. (London) A236, 119 (1958); F. G. Allen, J. Phys. Chem. Solids 19, 87 (1961).
9. N. P. Keier, S. Z. Roginskii, and I. S. Sazonova, Izvest. Akad. Nauk SSSR, Ser. Fiz. 21, 183 (1957).
10. E. Kh. Enikeev, L. Ya. Margolis, and S. Z. Roginskii, Doklady Akad. Nauk SSSR 130, 807 (1960).
11. J. J. Lander, J. Phys. Chem. Solids 15, 324 (1960).
12. Sh. M. Kogan and V. B. Sandomirskii, Doklady Akad. Nauk SSSR 127, 377 (1959).
13. V. M. Frolov, O. V. Krylov, and S. Z. Roginskii, Doklady Akad. Nauk SSSR 126, 107 (1959); V. M. Frolov, Dissertation for Candidate's Degree (Moscow, 1961).
14. J. Bardeen, Phys. Rev. 71, 717 (1947).
15. S. Z. Roginskii, Problemy Kinetiki i Kataliza Akad. Nauk. SSSR 8, 110 (1955).
16. A. N. Terenin, Problems of Photosynthesis (Izd. AN SSSR, 1959) p.9.
17. K. Hauffe, Bull. soc. chim. Belgrade 67, 417 (1958).
18. G. Tamman, Z. anorg. u. allgem. Chem. 149, 21 (1925).
19. A. B. Shekhter and I. I. Tret'yakov, in the collection: Heterogeneous Catalysis in Chemical Industry (Goskhimizdat, 1955) p.5.
20. I. A. Hedvall, Handbuch der Katalyse (Athens, 1943), Volume VI; Reaktionsfähigkeit fester Stoffe (Leipzig, 1938).
21. J. Bardeen and C. Herring, Imperfections in Nearly Perfect Crystals (New York and London, 1952) p.261; W. Jost, Diffusion und chemische Reaktion in festen Stoffen (1937).

22. A. B. Shekhter, A. I. Echeistova, and I. I. Tret'yakov, Zhur. Fiz. Khim. <u>24</u>, 202 (1950).

23. A. Mel'nikov, Fundamentals of Chemisorption (Oborongiz, 1938).

24. S. Z. Roginskii, Adsorption and Catalysis on Inhomogeneous Surfaces (Izd. AN SSSR, 1948).

25. S. Z. Roginskii, Zhur. Fiz. Khim. <u>31</u>, 2381 (1957).

26. S. Z. Roginskii, Zhur. Fiz. Khim. <u>32</u>, 737 (1958).

27. H. E. Farnsworth, Semiconductor Surface Physics (New York, 1957) p.3.

28. P. Handler, Semiconductor Surface Physics (New York, 1957) p.23.

29. V. A. Gordeeva, E. V. Egorov, G. M. Zhabrova, B. M. Kadenatsi, M. Ya. Kushnerev, and S. Z. Roginskii, Doklady Akad. Nauk SSSR <u>136</u>, 1364 (1961).

30. Radiation Effects in Solids (New York and London, 1957).

31. S. Z. Roginskii, Kinetika i Kataliz <u>1</u>, 15 (1960).

32. A. N. Terenin, Problemy Kinetiki i Kataliza Akad. Nauk SSSR <u>10</u>, 214 (1960).

33. B. P. Bering, Doctoral Dissertation (Moscow, 1957).

34. G. I. Chizhikova and N. P. Keier, Problemy Kinetiki i Kataliza Akad. Nauk SSSR <u>10</u>, 77 (1960).

35. S. Z. Roginskii, Doklady Acad. Nauk SSSR <u>126</u>, 817 (1959).

36. G. Parravano, H. G. Friederick, and M. Boudart, J. Phys. Chem. <u>63</u>, 1144 (1959).

37. V. I. Popov and S. Z. Roginskii, Kinetika i Kataliz <u>2</u>, 705 (1961).

38. Collection: Catalysis. Proceedings of the Third Conference on Physico-Chemical Problems, 1928 (Nauchnoe khim. tekhn. izd., Leningrad, 1930).

39. S. Z. Roginskii, Zhur. Fiz. Khim. <u>15</u>, 1 (1941).

40. M. Letort and P. le Goff, J. chim. phys. <u>53</u>, 480 (1956); <u>54</u>, 3 (1957).

41. M. I. Duell and A. J. B. Robertson, Trans. Faraday Soc. <u>57</u>, 1416 (1961).

42. H. W. Kohn and E. H. Taylor, J. Phys. Chem. <u>63</u>, 966 (1959); T. I. Barry and R. Roberts, Nature <u>184</u>, 1061 (1959).

43. F. Romero-Rossi and F. S. Stone, Second International Congress on Catalysis (Paris, 1960), Sec. II, Preprint 72.

44. G. K. Boreskov, in the collection: Heterogeneous Catalysis in the Chemical Industry (Goskhimizdat, 1955), p.5.

45. S. Z. Roginskii, I. I. Tret'yakov, and A. B. Shekhter, Zhur. Fiz. Khim. <u>29</u>, 1921 (1955).

46. K. Fricke, Handbuch der Katalyse (Athens, 1943) Vol. IV.

47. I. Müttig, Handbuch der Katalyse (Athens, 1943) Vol. IV.

48. K. Hauffe, Reaktionen in und an festen Stoffen (Berlin, 1955).

49. S. Z. Roginskii, Zhur. Fiz. Khim. <u>12</u>, 47 (1938); S. Z. Roginskii and O. M. Todes, Izvest. Akad. Nauk SSSR, Otdel. Khim. Nauk 475 (1940).

50. S. Z. Roginskii, in the collection: Heterogeneous Catalysis in Chemical Industry (Goskhimizdat, 1955).

51. P. Jacobs and F. C. Tompkins, in the collection: Chemistry of Solid State, ed. by W. E. Garner [Russian Translation] (IL, 1961) p.81.

52. V. V. Boldyrev, Zhur. Fiz. Khim. <u>33</u>, 2539 (1959); Kinetika i Kataliz <u>1</u>, 203 (1960).

53. J. Fleeman and G. J. Dienes, J. Appl. Phys. <u>26</u>, 652 (1955).

54. R. Hurst and J. Wright, Proc. Intern. Conf. on Peaceful Uses of Atomic Energy (United Nations, 1956) Paper No. 900.

55. P. B. Weisz and E. N. Swegler, J. Chem. Phys. <u>23</u>, 1567 (1955).

56. V. A. Kargin and V. A. Kabanov, International Symposium on Macromolecular Chemistry (Izd. AN SSSR, 1960), Sec. II, p.453; V. A. Kargin and N. A. Platé, ibid, p.460.

57. M. Maga, Khimiya i tekhnologiya polimerov No. 7-8, 102 (1960).

58. N. N. Semenov, ibid, p.196.

EFFECT OF OXYGEN AND WATER VAPOR ON THE SURFACE PROPERTIES OF GERMANIUM AND SILICON

R. Kh. Burshtein, L. A. Larin, and S. I. Sergeev

Electrochemistry Institute, Academy of Sciences, USSR

It is known that the state of the surface of germanium and silicon strongly affects their electrical properties. Therefore, studies of the mechanism of adsorption and of the properties of layers adsorbed on these semiconductors are of special interest.

Several papers have been published on the process of chemisorption of oxygen on germanium and silicon and work has been done on the influence of chemisorbed oxygen layers on the electron work function and the surface recombination velocity of these semiconductors.

However, the studies of their electrical properties have not been accompanied by parallel studies of the process of oxygen chemisorption and, therefore, the special features of the adsorption process, found in later work [1-4], have not been allowed for in measurements of the contact potential difference and the surface recombination velocity.

Moreover, the papers mentioned above dealt with surfaces subjected to different cleaning methods, which may have had considerable effect on the state of the surface.

From several published papers, it is known that the properties of semiconducting germanium devices are altered on contact with water vapor. In this connection, it was interesting to investigate, by adsorption and electrical methods, the effect of water vapor on the surface properties of germanium and silicon, and on the interaction of oxygen with these two semiconductors.

The kinetics of chemisorption and the influence of adsorbed oxygen and water vapor on the surface properties of germanium and silicon were studied in our laboratory. In these studies adsorption methods were used as well as methods employing the contact potential difference and the surface recombination velocity.

Since the surface of germanium or silicon is always covered with an oxide layer, it is important to develop a method of cleaning these semiconductors for the purpose of studying adsorption of gases. In recent work especial attention has been paid to this point. Farnsworth and co-workers [5], Handler [6], and others [7] used the ion bombardment method to clean the surface of germanium and silicon.

Law and Garrett [8] are of the opinion that the surface of germanium treated by ion bombardment is not free of chemisorbed oxygen. There are also indications that after ion bombardment the germanium surface has a large number of defects.

The method of obtaining a clean germanium surface by the condensation of films [9] is also unreliable because during the evaporation of germanium the growing film may capture the gases evolved on heating. A more reliable variant of this method is the technique developed very recently by Bennett and Tompkins [3].

This technique involves melting germanium in hydrogen, outgassing it and depositing it on a cool wall of a chamber. Some workers obtained clean surfaces for adsorption measurement by fragmenting single crystals in ultravacuum.

According to published data, the surface of a silicon single crystal also can be cleaned by ion bombardment or heating in vacuum.

To clean a semiconductor surface it is desirable to use a method which is convenient both for adsorption measurements and for studies of electrical properties.

In the present work we removed oxide layers from the surface of germanium by multiple reduction in hydrogen, both for investigations of oxygen chemisorption and for electrical measurements.

After the reduction at 400-450°C, a prolonged outgassing was carried out in 10^{-7} mm Hg vacuum at the same temperature. The reduction and the outgassing were repeated 5-6 times. In some experiments, for complete removal of hydrogen from the reduced germanium surface, a prolonged outgassing in 10^{-9} mm Hg vacuum at 400-450°C was carried out.

According to Tamaru and Boudart [10] the complete desorption of hydrogen from germanium is obtained even at 280°C. From the results of our experiments it also follows that outgassing at 400°C is capable of removing completely the adsorbed hydrogen from the surface of germanium.

This cleaning method is convenient because it is applicable both to single crystals and to powders. This allows us to start with the same initial surface in studies of the electrical properties of the surface and in studies of the chemisorption of oxygen on germanium. Other methods of surface cleaning are convenient only either for adsorption measurements or for electrical measurements.

The adsorption of oxygen was investigated using germanium powder prepared by crushing a single crystal and cleaning it by the method described above. The surface area of the powder, determined by the Brunauer, Emmett, and Teller (BET) method, was 620 cm^2/g. Adsorption tests were carried out using 60.4 g of germanium powder. The use of a large surface area allowed us to investigate the process of oxygen chemisorption in greater detail.

The adsorption of oxygen on germanium was studied by the method we employed earlier in investigating the adsorption of oxygen on iron [11]. This method was used also by Rideal and Trapnell [12] to study the adsorption of oxygen on tungsten.

In our experiment the germanium powder was cooled to room temperature after reduction and outgassing and then the oxygen was adsorbed in small portions and the rate of absorption of each portion was measured. The initial pressure of each oxygen portion was about 0.07 mm Hg. The results obtained for the adsorption of oxygen on germanium are given in Fig. 1. From these results it is clear that there are two stages of oxygen sorption in germanium: the fast and the slow. The rate of adsorption in the fast stage is practically independent of the degree of surface coverage. The rapidly adsorbed oxygen represents the formation of a monatomic layer on the assumption that one atom of germanium on the surface adsorbs one atom of oxygen. The number of atoms per 1 cm^2 of the germanium powder surface was taken to be 7.7×10^{14}, in accordance with the data of Green, Kafalas, and Robinson [1].

The fast stage of oxygen adsorption could be used to determine the surface area of germanium. The surface area of 630 cm^2 obtained for 1 g of germanium powder is in good agreement with the results obtained by the BET method. Figure 2 shows the dependence of $\log(1/\tau)$, on the amount of absorbed oxygen, where τ is the absorption time of half a portion of oxygen. This figure indicates that the amounts of slowly and rapidly absorbed oxygen are equal.

It follows from our experiments that the kinetics of the slow chemisorption stage is logarithmic: $N \approx \log t$, where N is the amount of adsorbed oxygen and t is the time from the beginning of the adsorption process. This result is in agreement with the data of Green, Kafalas, and Robinson [1], according to whom the fast stage of oxygen chemisorption on a surface prepared by crushing a single crystal in ultravacuum, represents the formation of 0.9 monolayer. In his later work, Tompkins has shown that the amount of oxygen chemisorbed

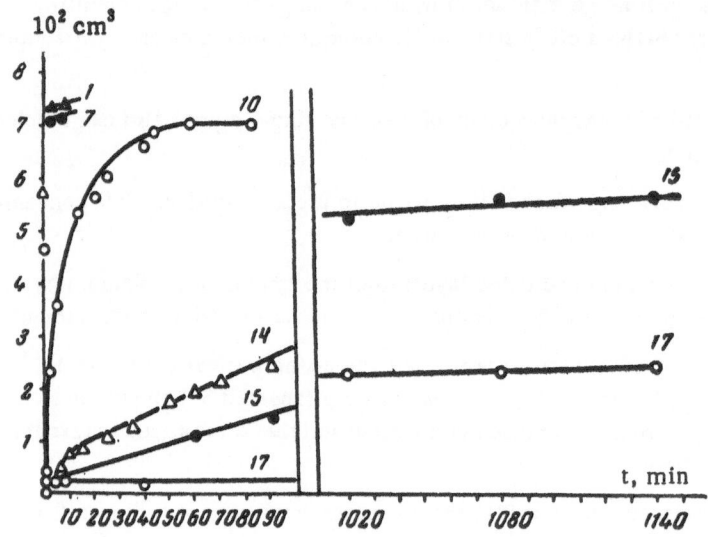

Fig. 1. Kinetics of the adsorption of separate portions of oxygen on germanium (the numbers by the curves give the order in which the oxygen portions were admitted).

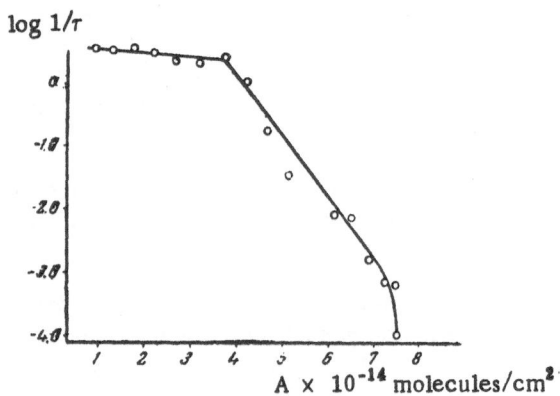

Fig. 2. Dependence of log $1/\tau$, where τ is the time necessary to absorb half a portion of oxygen, on the number of adsorbed oxygen molecules per 1 cm² of the surface.

in the fast stage increases from 0.3 to 1 monolayer as the temperature is raised from -78 to 0°C. And according to Brennan, Hayward, and Trapnell [4], the fast chemisorption stage represents a surface coverage of 0.83 monolayer; they found that the heat of adsorption in this range of surface coverage decreases from 123 to 100 kcal/mole. These data are obviously more accurate than those of Green et al. Thus both in our measurements and on surfaces cleaned by the other aforementioned methods, the fast chemisorption stage represents the formation of between 0.83 and 1 monatomic layer on the assumption that one surface atom of germanium adsorbs one atom of oxygen.

In the slow chemisorption stage, similar values are obtained for the amount of chemisorbed oxygen, representing about 1 monolayer for surfaces cleaned by our method and by the method of fragmentation of a single crystal in ultravacuum. Comparison of these various adsorption data is an additional verification of the cleanness of the surface used in our experiments.

The feasibility of the removal of oxide layers from the germanium surface by vacuum heating was investigated experimentally. For this purpose, the germanium powder, on which the fast and slow stages of adsorption were completed at room temperature, was heated at 400°C in vacuum for 2 hours. No desorption of oxygen was observed on heating. After cooling the powder, its ability to adsorb oxygen at room temperature was determined. It was found that after heating the germanium powder, the slow stage of adsorption occurred again.

Since the desorption of oxygen does not occur on outgassing, and there is practically no removal of oxygen from the surface in the form of GeO [13], we may assume that, on heating germanium on which oxygen was chemisorbed, the nature of the bonding between oxygen and the surface atoms of germanium is altered.

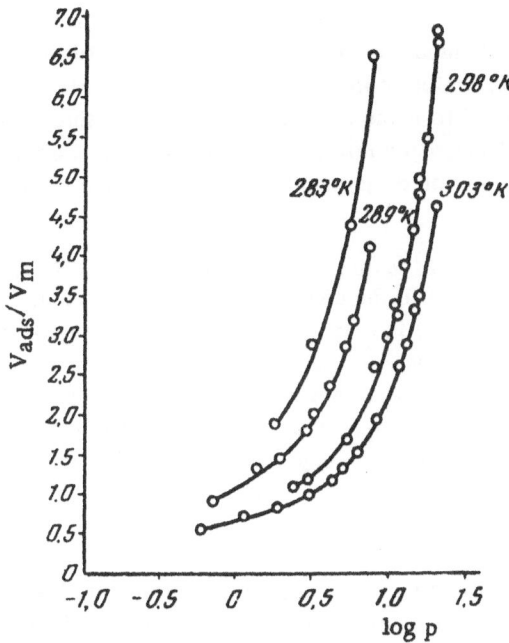

Fig. 3. Isotherm of the water vapor adsorption on the surface of germanium according to Law [14].

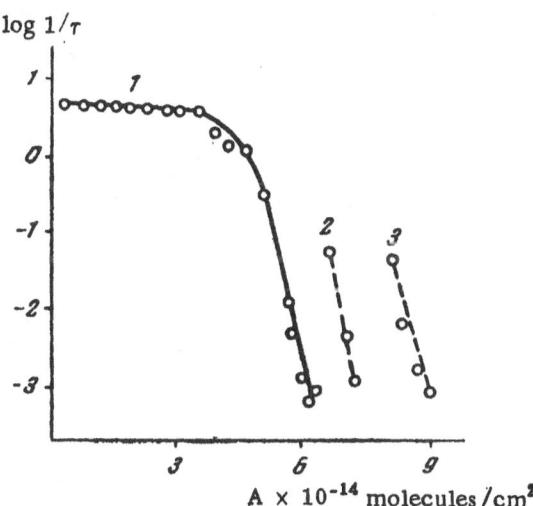

Fig. 4. Dependence of the oxygen adsorption rate on the surface coverage of germanium: 1) clean surface; 2) and 3) after adsorption of water vapor.

The adsorption of water vapor on the surface of germanium was investigated by Law [14] and Kawasaki et al. [15]. From their results it follows that, with the increase of the relative humidity to 80%, the amount of water vapor adsorbed on germanium increases to 3 molecular layers, and on further increasing the relative humidity to 100% the amount of adsorbed water increases to 7 molecular layers. The results of Law are given in Fig. 3.

The adsorption of water vapor on a cleaned surface of germanium is in practice reversible, as indicated by the measurements of the work function.

Preliminary tests on the adsorption of water vapor on an oxidized surface showed that in the presence of this vapor some irreversible changes of the surface properties of germanium take place. In connection with this, the process of the influence of water vapor on the interaction of germanium and oxygen was studied in detail. The adsorption tests were carried out as follows. On a clean surface of germanium freed from the oxide layer by reduction in hydrogen followed by outgassing (as described earlier), oxygen was adsorbed in small portions sufficient for completion of the fast and slow stages of adsorption. Then the germanium surface was placed in contact with water vapor. After some time the water vapor was frozen out or pumped out and again the kinetics of oxygen sorption was investigated.

The results of these tests are shown in Fig. 4, where the ordinate represents the reciprocal of the time necessary to adsorb half the oxygen in a given portion and the abscissa represents the total amount of oxygen adsorbed by the germanium surface. From these results it is clear that, if, after completion of the fast and slow adsorption processes represented by curve 1, the germanium is placed in contact with water vapor, then the rate of adsorption of O_2 increases after the removal of the water vapor [16]. As fresh portions of oxygen are adsorbed, the rate of adsorption decreases (Fig. 4, curve 2). However, a subsequent interaction of the germanium, covered anew by a passivating oxide layer, with water vapor again increases the rate of the germanium-oxygen interaction, but in this case the passivation of the surface may be reached by additional sorption of oxygen (Fig. 4, curve 3). Thus these tests show that on the adsorption of water vapor on a germanium surface coated with an oxide layer the passivating properties of the oxide layer are disturbed, which leads to formation of a thick oxygen layer on the surface if oxygen is present.

To find the effect of the various stages of oxygen chemisorption on the surface properties, we investigated the influence of adsorbed oxygen on the electron work function of germanium [17].

The contact potential difference was measured in the present work by the Kelvin vibrating capacitor method. The main disadvantage of this method is the adsorption of gases both on the test electrode as well as on the reference electrode, which makes it difficult to obtain correct results on the influence of adsorbed gases on the electron work function. In order to avoid this difficulty, we used a device developed earlier [18] in which the reference electrode is coated with glass. Since the adsorption of many gases on glass is considerably smaller than on metals, such a reference electrode is more stable. To avoid interference during measurements, the device was screened after heating at 400°C. The use of a glass-coated reference electrode made it possible to study the influence of adsorbed oxygen on the electron work function of germanium in a wide range of temperatures.

The electron work function of both n- and p-type germanium was 0.73-0.75 eV smaller than the work function of the reference electrode. The absence of a large difference in the electron work function (±0.02 eV) in spite of the considerable difference in the position of the Fermi level in n-type and p-type germanium is due to the high density of surface states [19]. The results obtained in the study of variation of the contact potential difference of a germanium surface during the adsorption of oxygen at various pressures, are given in Fig. 5. These results refer to an n-type germanium sample of ρ = 20 $\Omega \cdot$cm resistivity. Similar (within the limits of ±0.02 eV) results were obtained for other samples of n- and p-type germanium having various conductivities.

The influence of adsorbed oxygen on the work function was investigated in the pressure range 10^{-3}-10^2 mm Hg. At each pressure the contact potential difference was measured over an interval of 20 min.

From the results obtained, it follows that the electron work function of n- and p-type germanium increases with the adsorption of oxygen. The dependence of the contact potential difference on the logarithm of oxygen pressure is linear over a wide range of pressures.

Comparison of the results on the work function with the results of the adsorption measurements indicated that the fast stage of oxygen adsorption on germanium changes the contact potential difference by 0.10-0.15 V. This fast stage represents the formation of a monatomic layer and ends after not more than 5 min even at a pressure of 10^{-3} mm Hg. On further increase of the oxygen pressure and the consequent increase of the amount of adsorbed oxygen, corresponding to the slow stage, the work function continued to rise and increased by 0.48 eV at 100 mm Hg. The results obtained do not agree with the data of Dillon [20], who found that the maximum increase of the electron work function by 0.2 eV occurred at the oxygen pressure of p = 10^{-6} mm Hg; for p_{O_2} = 10^{-5} mm Hg Dillon reported that the electron work function decreased by 0.06 eV and that further increase of the pressure did not affect the work function. According to Dillon and Farnsworth [21] the work function increases by 0.2 eV on the adsorption of oxygen. Dillon and Farnsworth believed that the contact potential was practically unaffected by pressure variation. However, these workers varied the pressure only from 1 x 10^{-7} to 2 x 10^{-5} mm Hg. Under these conditions the slow stage of adsorption is difficult to observe.

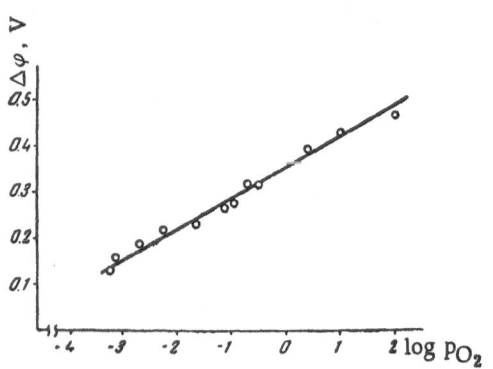

Fig. 5. Dependence of the contact potential difference between germanium and a reference electrode on the logarithm of oxygen pressure.

According to our results, several days are needed to complete the slow stage of oxygen chemisorption at 0.07 mm Hg. On increasing the pressure the rate of adsorption in the slow stage increases as $p^{0.52}$ [1, 3]. This may explain the variation of the contact potential difference as the pressure is increased. The change in the contact potential difference on completion of the fast stage is smaller than that produced in the slow stage. The results obtained lead us to conclude that during the chemisorption of the second layer of oxygen atoms on the germanium surface a new compound is formed. If it is assumed that the fast chemisorption stage corresponds to a reaction of the type

$$Ge + \frac{1}{2} O_2 \rightarrow GeO, \qquad (1)$$

then the slow adsorption stage is apparently related to

formation of a layer of the GeO_2 type by the reaction

$$GeO + \frac{1}{2} O_2 \rightarrow GeO_2.$$ (2)

From the results obtained by measuring the contact potential difference, it follows that at pressures up to 10 mm Hg oxygen is adsorbed irreversibly on the germanium surface. This is indicated by the fact that outgassing to 10^{-6} mm Hg after the adsorption of oxygen does not alter the contact potential difference. However, in those cases where the influence of adsorbed oxygen on the contact potential difference was studied at pressures of 100 mm Hg or more, there was also, apart from the irreversible adsorption, the reversible adsorption which has some effect on the contact potential difference. Thus, for example, after adsorption at p_{O_2} = 100 mm Hg outgassing reduced the electron work function by 0.04–0.05 eV, which may possibly have been due to physical adsorption of oxygen on the surface of the oxide.

In order to investigate the behavior of oxygen chemisorbed on germanium at various temperatures, a definite amount of oxygen was adsorbed on germanium at room temperature. The contact potential difference between germanium and the reference electrode was measured, and then germanium with chemisorbed oxygen was heated at various temperatures in gaseous media free of oxygen. After heating for 1 hour, the device in which germanium was placed was cooled to room temperature and the contact potential difference was measured again. The results obtained in this study of the effect of the temperature of heating of germanium with chemisorbed oxygen on the electron work function, are given in Fig. 6. From these results it follows that after heating at temperature of 100–400°C the electron work function of germanium with a layer of chemisorbed oxygen decreases (Fig. 6, curve 1), which may be explained either by penetration of oxygen into germanium or the evaporation of germanium oxide from the surface or the decomposition of the oxide. On heating to 200°C, the work function is reduced by 0.2 eV. After heating to 400°C, the work function is only 0.08 eV higher than that of a clean germanium surface.

These results are in agreement with those of Schlier and Farnsworth [5] who showed, by electron diffraction, that heating at 500°C in vacuum seemed to have cleaned a surface of germanium on which oxygen was chemisorbed. Our adsorption measurements showed that, after heating to 400°C the germanium on which the fast and slow adsorption stages were completed, oxygen was again adsorbed slowly at room temperature. Since the desorption of oxygen did not occur on heating, we may assume that heating produced the reaction

$$GeO_2 + Ge \rightarrow GeO.$$ (3)

The GeO formed on contact with oxygen at room temperature was again transformed into a compound of the GeO_2 type.

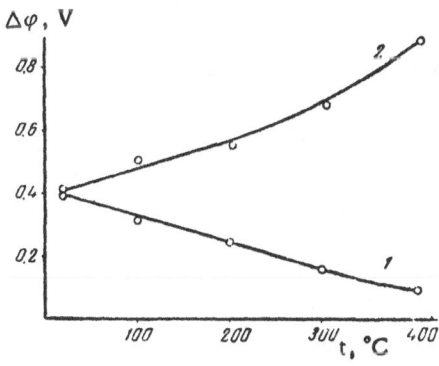

Fig. 6. Dependence of the contact potential of a germanium surface with chemisorbed oxygen on the heating temperature: 1) in 1×10^{-6} mm Hg vacuum; 2) in oxygen, p_{O_2} = 5 mm Hg.

Thus the heating of germanium to 400°C in vacuum did not remove the chemisorbed oxygen from the surface but altered the nature of the binding of the oxygen and germanium. On this basis the change in the contact potential difference can be explained by the reaction (3).

Different results were obtained on heating germanium in the presence of some oxygen in the gaseous phase (5 mm Hg). These results, shown by curve 2 in Fig. 6, indicate that as the heating temperature is increased (after preliminary heating at 400°C) the electron work function of germanium increases by 0.9 eV which suggests an increase of the thickness of the GeO_2 layer.

Thus it follows from our experiments that the electron work function of germanium continues to change even when there are several oxide layers on the surface. These results are in agreement with our earlier data on the adsorption of oxygen on iron and they may be explained by

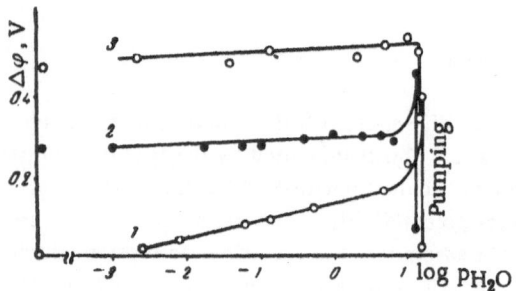

Fig. 7. Dependence of the contact potential difference on log p_{H_2O}: 1) clean surface; 2) germanium covered with an oxide layer; 3) after multiple treatment.

Mott's representation of the diffusion of metal atoms [22], in our case semiconductor atoms, to the surface of the oxide.

The conclusion, drawn above from the adsorption measurements, that the protective properties of oxide layers are disturbed in the presence of water vapor is very clearly confirmed by measurements of the contact potential difference [16].

The results on the influence of water vapor, adsorbed by a clean germanium surface, on the electron work function are given by curve 1 in Fig. 7. They show that as the water vapor pressure is increased the work function increases linearly with log (p_{H_2O}) in the pressure range from 1×10^{-3} to 7 mm Hg. In this range of pressures, the electron work function increases by 0.16 eV. However, on a further increase of the water vapor pressure (corresponding to an increase of the relative humidity from 50 to 100%) the electron work function increases more markedly (by 0.24 eV). Thus the total increase of the work function on the adsorption of water vapor amounts to 0.4 eV. This change of the electron work function is reversible. On pumping out the water vapor, the work function assumes a value close to that of a clean germanium surface.

The influence of water vapor on the electron work function of germanium covered with an oxide layer is different. The results of the relevant tests are given by curve 2 in Fig. 7. In these tests the oxygen was adsorbed on germanium for 15 hours at a pressure of 0.1 mm Hg. According to the earlier results the contact potential change under these conditions is due to the fast adsorption of oxygen on germanium (a change by 0.15 V) and partly due to the slow adsorption. As shown by curve 2 in Fig. 7 the increase of the work function due to chemisorbed oxygen is, under these conditions, 0.28 eV. The adsorption of water vapor on a germanium surface covered by an oxide layer changes the contact potential by only 0.02 V for a change of pressure from 1×10^{-3} to 7 mm Hg. On increasing the water vapor pressure from 7 to 15 mm Hg, the contact potential difference shifts by 0.16 V. Thus the adsorption of water vapor on the surface of germanium covered with an oxide layer produces a contact potential shift of about half that on a clean surface.

Curves 1 and 2 in Fig. 7 indicate that in the 7-15 mm Hg range of the water vapor pressure (log p_{H_2O} = 0.84-1.18) there is a strong variation of the contact potential difference with increasing pressure. The results obtained may be explained on the basis of the work of Law and of Kawasaki et al. [14, 15], from which it follows that with the increase of the relative humidity to 80% the amount of adsorbed water vapor represents

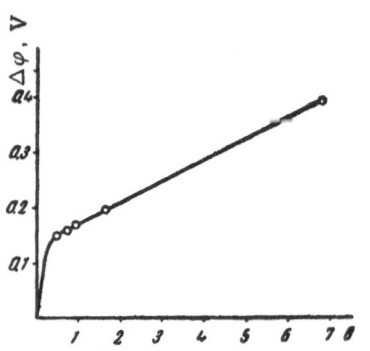

Fig. 8. Dependence of the contact potential difference on the number of adsorbed water molecules; θ is the number of molecular layers.

the formation of three molecular layers and that with a further increase of the relative humidity from 80 to 100% the water layer thickness increases to seven molecular layers. Comparison of our results with those of the Japanese [15] is given in Fig. 8, which shows the dependence of $\Delta\varphi$ on the number of molecular layers of water adsorbed on germanium. It is seen in Fig. 8 that a considerable increase of the contact potential difference at higher pressures is due to a considerable increase of the adsorption of water vapor under these conditions.

In contrast to the adsorption of water vapor on a clean surface, the adsorption of this vapor on a surface of germanium covered with an oxide layer produces irreversible changes of the surface properties (curve 2 in Fig. 7).

This follows from the fact that after the desorption of water vapor the contact potential difference does not return to the value before adsorption of the vapor, but remains 0.07 V higher than the contact potential difference of a clean surface of germanium, which indicates that the structure of the oxide layer is disturbed and its protective properties are altered.

It is possible that the smaller change of the contact potential difference produced by the action of water vapor on an oxidized surface of germanium, compared with the effect on a clean surface, is related to the fact that in addition to the increase of the work function occurring on the adsorption of water vapor there is also the reduction of the work function due to the disturbance of the oxide layer structure.

According to the data in Fig. 4, this disturbance of the structure makes a germanium surface capable of additional sorption of oxygen. After additional adsorption of oxygen the value of the contact potential difference increases to 0.4 eV. The process of the consecutive action of oxygen, water vapor and vacuum on the surface properties of germanium was followed for 12 cycles. The outcome of these tests is shown in Fig. 9. Unless indicated otherwise in Fig. 9, oxygen was adsorbed on germanium at a pressure of the order of 0.1 mm Hg for 15-30 min, and water at a relative humidity of 100% for 10-30 min. The sample was kept in vacuum for about 30 min. From the results in Figs. 9 and 4, it is clear that in the presence of water vapor it is impossible to reach a stable state of the germanium surface; as indicated by the results on the value of the work function, after the desorption of water vapor, every time that the oxidized surface is placed in contact again with water vapor the protective properties of the oxide layer are destroyed.

The results obtained in a study of the effect of water vapor at various pressures on the contact potential difference after many treatments with oxygen and water vapor (11 cycles) are given by curve 3 of Fig. 7. Thus it follows that when a thick oxide layer is formed on a germanium surface the adsorption of water vapor even at a high relative humidity increases the work function by only 0.03 eV, which is considerably less than the change of the work function in the first cycle. However, on pumping the water vapor out, the work function continues to change by a considerable amount, although this change is smaller than in the first cycle: the work function changes by 0.4 eV on evacuation of the water vapor in the first cycle, but in the tenth cycle this change is 0.16 eV.

An investigation of the simultaneous action of oxygen and water vapor on the electron work function of germanium was carried out by Brattain and Bardeen [23]. In their experiments on germanium treated with oxygen in a spark discharge, the electron work function decreased both in an atmosphere of oxygen and in an atmosphere of oxygen saturated with water vapor. The results of our experiments indicate that the adsorption of oxygen and the adsorption of water vapor on a clean germanium surface and on a surface covered by a not too

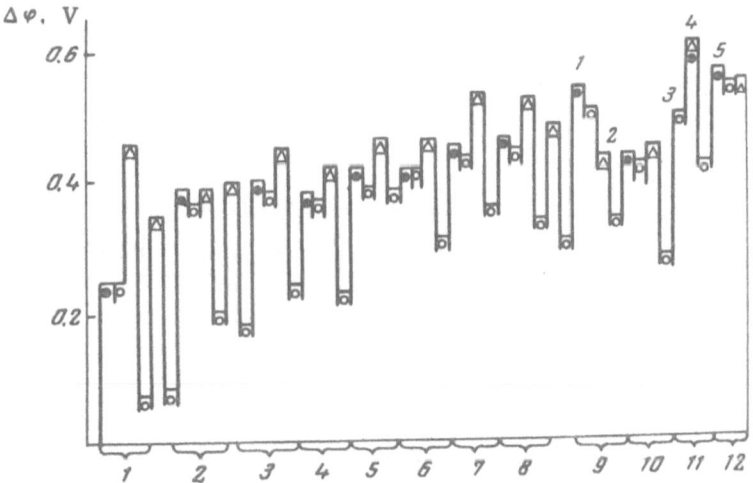

Fig. 9. Variation of the contact potential difference on repeated treatment of a germanium surface with oxygen, water vapor and vacuum; ● - oxygen; △ - water; ○ - vacuum. 1) p_{O_2} = 12 mm Hg; 2) τ = 120 min; 3) p_{O_2} = 3 mm Hg; 4) simultaneous action of water vapor (15 mm Hg) and oxygen (3 mm Hg); 5) p_{O_2} = 1.5 mm Hg, τ = 1000 min.

thick oxide layer produces an increase of the electron work function. Obviously the difference between the two sets of results is related to the fact that treatment of germanium in a spark discharge in oxygen essentially alters the surface properties of germanium.

From our results, it follows that the adsorption of water vapor destroys the protective oxide layer on the germanium surface. This in turn leads to the formation of thick oxide layers on the surface. It is possible that the deterioration of the parameters of semiconducting devices in contact with the atmosphere is due to the formation of such layers.

It was interesting to find how various forms of chemisorbed oxygen affect the lifetime of minority carriers in germanium.

To investigate this problem the minority-carrier lifetime was studied on a clean germanium surface and on a surface which chemisorbed oxygen.

The influence of oxygen on the carrier lifetime, determined from the photoconductivity decay, was investigated by Madden and Farnsworth [24]. The germanium surface in their tests was subjected to ion bombardment followed by heating in vacuum; the oxygen pressure did not exceed 1.2×10^{-4} mm Hg. After such treatment the oxygen adsorbed at room temperature did not affect the carrier lifetime; but the oxygen adsorbed at 100°C reduced the minority-carrier lifetime on the germanium surface. One should note that in Madden and Farnsworth's work the tests were carried out at very low oxygen pressures, which should have reduced the chemisorption rate, particularly in the slow stage. To compare the results on the carrier lifetime with those on the chemisorption kinetics and the electron work function, it was necessary to carry out measurements of the lifetime under the same conditions as in our earlier tests.

The minority carrier lifetime was measured by the photogalvanomagnetic method [25, 26]. An electromagnet (B = 3200 G) provided the magnetic field. A 500 W projection lamp was used as the light source, and the light was modulated at 60 cps. To amplify the signal, a narrow-band amplifier with a gain of 6×10^4 was employed.

A rectangular germanium slab was placed in a special holder in a device shown in Fig. 10. The end contacts were used to pass direct current through the sample and to measure the photogalvanomagnetic effect and photoconductivity emf's. The temperature was measured with a thermocouple placed near the surface of germanium.

The tests were carried out on n-type germanium samples having $\rho = 20$ $\Omega \cdot$cm, L = 1.5 mm, and $\rho = 48$ $\Omega \cdot$cm, L = 3.2 mm.

The samples were ground and then etched in hydrogen peroxide.

The surface was cleaned by multiple reduction in hydrogen at 400°C followed by outgassing in 10^{-7} mm Hg at the same temperature. The carrier lifetime in germanium samples treated in this way was practically the same as on a freshly etched surface.

All measurements of the carrier lifetime were carried out at room temperature. A study was made of the effect, on the lifetime, of oxygen chemisorbed at various pressures in the temperature range 20-400°C. The results obtained at 20°C are given in Fig. 11, where the abscissa represents the logarithm of the oxygen pressure and the ordinate shows the effective lifetime. The lifetime was measured 10 min after the admission of gas.

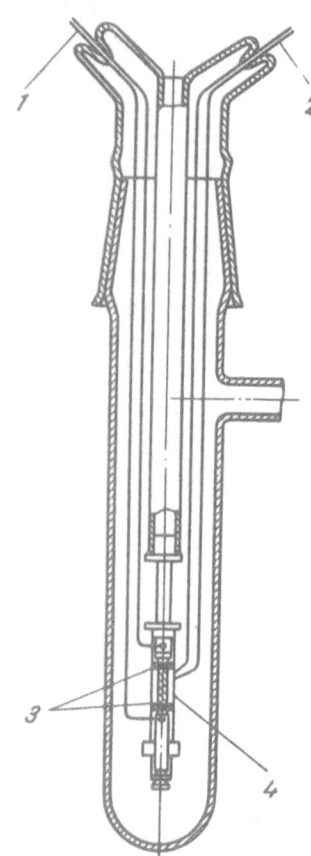

Fig. 10. Measuring device.
1) Lead to amplifier; 2) to pyrometer; 3) end contacts; 4) germanium sample.

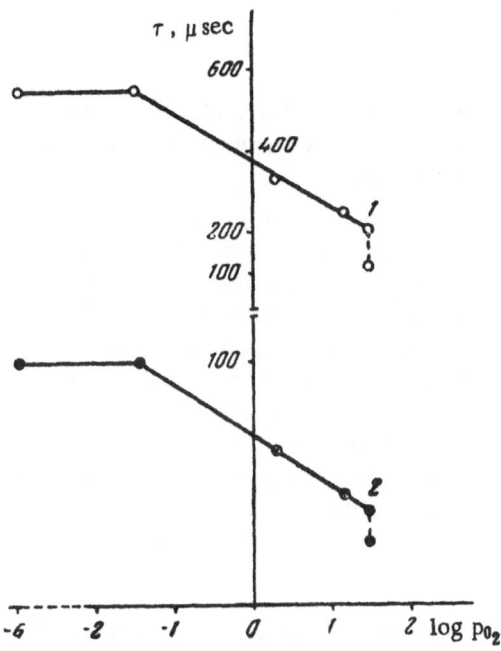

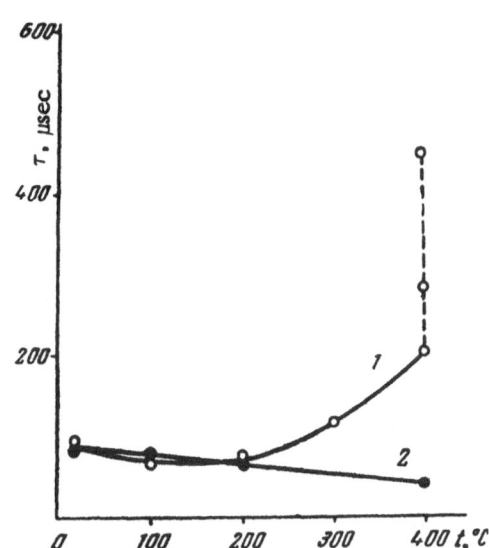

Fig. 11. Dependence of τ on log p_{O_2}: 1) n-type germanium, $\rho = 48\ \Omega \cdot cm$; 2) n-type germanium, $\rho = 20\ \Omega \cdot cm$.

Fig. 12. Dependence of τ on the heating temperature: 1) in vacuum; 2) in oxygen.

From the results obtained it follows that the fast and slow stages of chemisorption affect in different ways the minority-carrier lifetime on the surface of germanium. From the data in Fig. 11, obtained after 5 min from the beginning of a test at a given pressure, it is clear that at oxygen pressures not exceeding 1 mm Hg the chemisorption of oxygen does not affect the lifetime. Under these conditions, mainly the fast stage of chemisorption takes place. At higher pressures corresponding to the slow adsorption, the lifetime decreases and this decrease depends on the duration of contact of germanium with oxygen. It is clear from Fig. 11 that after a long time in oxygen (17 hours), during which the slow adsorption increases, the reduction of the lifetime τ corresponds to the dashed parts of the curves.

A study was also made of the effect of 20-400°C heating in vacuum on germanium with chemisorbed oxygen. These experiments were carried out on n-type germanium samples having $\rho = 48\ \Omega \cdot cm$, L = 3.2 mm. The results are given by curve 1 in Fig. 12. At each temperature germanium was heated for 1 hour. It is clear that with increase of the temperature of the heat treatment the lifetime increased. This increase was considerable at 400°C and the value of the lifetime after heating depended on the duration of heating. Heating in vacuum for 3 hours increased the lifetime to a value corresponding to a clean surface. An investigation was made also of the effect of oxygen adsorbed at high temperature on the lifetime of minority carriers. This showed that after the heating of germanium in the presence of gaseous oxygen (5 mm Hg), the lifetime decreased from 100 μsec at room temperature to 40 μsec after heating at 400°C (curve 2 in Fig. 12).

Comparison of the influence of the adsorbed oxygen on the work function and on the carrier lifetime allows us to conclude that the fast adsorption of oxygen, which does not affect the lifetime, produces a small change of the electron work function, while the slow chemisorption, during which a GeO_2-type layer is formed on the surface, considerably raises the work function and simultaneously strongly reduces the lifetime.

On vacuum heating of germanium with chemisorbed oxygen, i.e., on heating under conditions when the reaction (3) takes place, the work function decreases and the lifetime increases.

Heating of germanium in oxygen increases the work function and considerably reduces the lifetime. It has not yet been possible to draw a definite conclusion to what extent the results satisfy the theory of Garrett and Brattain [27], which established a relationship between the surface recombination velocity and the surface charge.

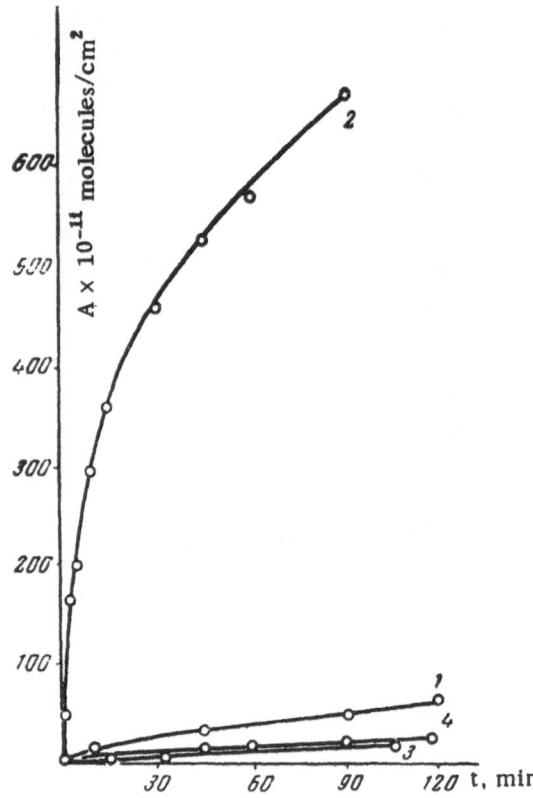

Fig. 13. Kinetics of oxygen adsorption on germanium. 1) Oxidized germanium surface; 2) after treatment with water vapor; 3) surface made hydrophobic; 4) hydrophobic surface after treatment with water vapor.

The results suggest that the change in the lifetime on adsorption of oxygen is related to a considerable degree to the formation of a GeO_2-type oxide on the germanium surface. The parts covered by this oxide are obviously surface recombination centers.

The mechanism of the influence of water vapor on the interaction of germanium with oxygen, described above, suggests that in order to prevent the formation of thick oxide layers on the surface of germanium under the influence of water vapor, it is desirable to make the surface hydrophobic. For this purpose trichlormethylsilane vapor was adsorbed on the surface after the fast and slow stages of oxygen adsorption were completed. Then trichlormethylsilane was hydrolyzed and polymerized at 150°C for 1.5 hours in the presence of water vapor.

The results on the kinetics of oxygen adsorption and the influence of water vapor on this adsorption are given in Fig. 13, where the ordinate represents the amount of adsorbed oxygen and the abscissa shows the time in minutes. Curve 1 represents the adsorption of oxygen on the oxidized germanium surface. Curve 2 shows how the kinetics of oxygen adsorption on the same germanium surface is affected by water vapor. Comparison of the results represented by curves 1 and 2 shows that the amount of oxygen chemisorbed on the germanium surface treated with water vapor exceeds by a factor of more than 130 the amount of oxygen chemisorbed during the same time (90 min) on the original surface. On the germanium surface which was made hydrophobic, the rate of adsorption after water-vapor treatment (curve 4) was practically the same as the rate of adsorption before water-vapor treatment (curve 3). Thus the activating effect of water vapor on the process of oxygen adsorption on germanium practically disappeared when the germanium surface was made hydrophobic. The stability of the hydrophobic surface with respect to oxygen adsorption was retained even after multiple treatments with water vapor and oxygen.

Experiments were also carried out to find the influence of adsorbed oxygen and water vapor on the surface properties of silicon.

To free the silicon surface from oxides we used heating in hydrogen at temperatures from 500 to 900°C followed by outgassing in 10^{-7} mm Hg vacuum at the same temperatures. The maximum duration of heat treatment was 100 hours.

According to Farnsworth the electrical properties of silicon surfaces treated in this way are close to properties of a surface subjected to ion bombardment.

A study of the interaction of oxygen with silicon showed that at room temperature there are two oxygen chemisorption stages (a fast one and a slow one), as in the case of germanium.

The results of one of the experiments, in which silicon powder was heated for 10 hours at 500°C, are given in Fig. 14, where the abscissa represents quantities proportional to the rate of oxygen adsorption (τ is the

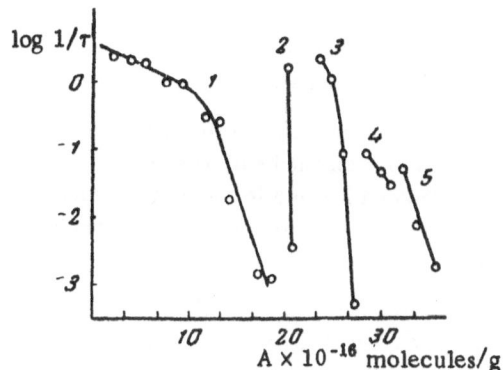

Fig. 14. Dependence of the oxygen adsorption rate on the surface coverage of germanium: 1) clean surface; 2) and 3) after adsorption of water vapor; 4) after heating at 200°C; 5) after heating at 500°C.

time for sorption for half a given portion of oxygen) and the ordinate shows the number of adso bed oxygen molecules per 1 g of silicon powder which had a specific surface area of 1200 cm²/g.

The amount of rapidly adsorbed oxygen corresponded to 12×10^{16} molecules/g. After longer heating (100 hours) the amount of rapidly adsorbed oxygen increased to 20×10^{16} molecules/g. Comparison of the results obtained on adsorption of oxygen with the value of the surface area, determined by the BET method, indicates that under these conditions only half the silicon surface is cleaned. The rate of chemisorption in the fast stage depends little on the surface coverage. The rate of oxygen chemisorption at the end of the fast stage decreases by a factor of about 5-7 compared with that at the beginning of this stage. On the other hand, the rate of oxygen chemisorption at the end of the slow stage is 1000 or more times lower than at the beginning (curve 1 in Fig. 14).

At 0.1 mm Hg the fast stage ends in several minutes. Several days are required to complete the slow stage at the same pressure.

The results of adsorption measurements showed that the amount of chemisorbed oxygen depends on the conditions of preliminary treatment of silicon powder.

On increase of the temperature of vacuum heating in a quartz ampoule from 500 to 700°C (for the same duration of heating) the amount of oxygen chemisorbed on silicon increased. However, further increase of the heating temperature to 900°C poisoned the surface and the amount of chemisorbed oxygen was considerably reduced.

Heating of silicon, on which the fast and slow stages of oxygen chemisorption had been completed, for one hour at 200°C in vacuum produced (curve 4 in Fig. 14) a small additional chemisorption of oxygen. Repeated heating at a higher temperature (500°C) for 2 hours increased the adsorption of oxygen by an additional amount (curve 5 in Fig. 14).

The effect of water vapor on the oxidized surface of silicon at 20°C made silicon capable of sorbing further amounts of oxygen (curves 2 and 3 in Fig. 14).

An investigation of the influence of the adsorbed oxygen on the electron work function of silicon showed that this oxygen increases the work function: the increase corresponding to the fast adsorption stage amounts to 0.12-0.15 V, while the oxygen adsorbed in the slow stage increases the work function by a further 0.6 V. Thus the slow adsorption stage has a considerable influence on the surface charge on silicon. In the slow stage of oxygen adsorption on silicon, as in the case of germanium, there is a semilogarithmic dependence of the change in the work function on the oxygen pressure.

Heating of the oxidized silicon surface at 300 and 450°C for several hours in vacuum strongly reduced the work function (curve 1 in Fig. 15) making it approach the value for a clean surface. Such an effect can be explained, in full accord with the results given earlier for the adsorption measurements, by the appearance of silicon atoms on the oxidized surface of silicon either due to the penetration of oxygen into the lattice or the emergence of silicon atoms on the surface as the result of vacuum heating. A similar change of the structure of the oxide layer on the surface of silicon in the presence of water vapor can be used to explain the change in the electron work function and the additional adsorption of oxygen on silicon after the action of water vapor on an oxidized silicon surface.

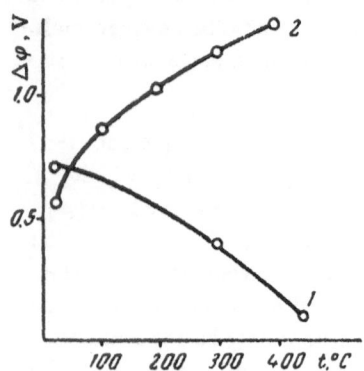

Fig. 15. Dependence of the contact potential of a germanium surface with chemisorbed oxygen on the heating temperature: 1) in 1×10^{-6} mm Hg; 2) in oxygen, 3 mm Hg.

The results obtained in measurement of the contact potential difference of silicon heated in oxygen at 3 mm Hg in the temperature range 100-400°C are given by curve 2 in Fig. 15.

After heating at 400°C the work function was 1.28 eV greater than the work function of a cleaned silicon surface. The increase of the work function on heating is obviously related, as in the case of germanium, to the formation of a thicker oxide layer on the silicon surface.

From the above results it follows that the fast and slow stages of the chemisorption affect in very different ways the surface properties of silicon and germanium. While the fast chemisorption affects little the work function and practically not at all the lifetime of carriers in germanium, the slow chemisorption affects the work function more strongly and produces a large change in the lifetime.

The adsorption and contact potential methods showed that in the presence of water vapor the protective properties of the oxide layer on the surface of germanium are disturbed and as a result of this a thick oxide layer is formed on the surface, which alters considerably the electrical properties of this semiconductor.

The disturbance of the protective oxide layer by water vapor may be due to the tendency of crystallization of the surface oxide or the penetration of semiconductor atoms to the surface of the oxide in the presence of water vapor.

The results obtained for germanium and silicon, together with the data for iron [29], suggest that it is more likely that water vapor promotes the penetration of semiconductor atoms to the oxide surface. This is supported by the agreement between the results on the effects of water vapor and of vacuum heating on an oxidized surface. Both these treatments produced the same changes in the surface properties, which are manifest as the reduction of the work function of the oxidized surface and as the ability to adsorb additional amounts of oxygen.

LITERATURE CITED

1. M. Green, J. A. Kafalas, and P. H. Robinson, Semiconductor Surface Physics (New York, 1956), p. 349.
2. R. Kh. Burshtein, L. A. Larin, and G. F. Voronina, Doklady Akad. Nauk SSSR 130, 801 (1960).
3. M. J. Bennett and F. C. Tompkins, Proc. Roy. Soc. (London) 259, 28 (1960).
4. D. Brennan, D. O. Hayward, and B. M. W. Trapnell, J. Phys. Chem. Solids 14, 117 (1960).
5. R. E. Schlier and H. E. Farnsworth, Semiconductor Surface Physics (New York, 1956), p. 3.
6. P. Handler, Semiconductor Surface Physics (New York, 1956), p. 23.
7. S. P. Wolsky and A. B. Fowler, Semiconductor Surface Physics (New York, 1957), p. 401.
8. J. T. Law and C. G. B. Garrett, J. Appl. Phys. 27, 656 (1956).
9. J. A. Allen, Revs. Pure and Appl. Chem. 4, 133 (1954).
10. K. Tamaru and M. Boudart, Advances in Catalysis 9, 699 (1957).
11. R. Kh. Burshtein, N. A. Shumilova, and K. A. Gol'berg, Zhur. Fiz. Khim. 20, 789 (1946).
12. E. K. Rideal and B. M. W. Trapnell, Proc. Roy. Soc. (London) A205, 409 (1951).
13. S. P. Wolsky, J. Appl. Phys. 29, 1132 (1958).
14. J. T. Law, J. Phys. Chem. 59, 67 (1955).
15. K. Kawasaki, K. Kanou, and Y. Sekita, J. Phys. Soc. Japan 14, 233 (1959).
16. R. Kh. Burshtein, L. A. Larin, and G. F. Voronina, Doklady Akad. Nauk SSSR 133, 148 (1960).
17. R. Kh. Burshtein and L. A. Larin, Doklady Akad. Nauk. SSSR 130, 565 (1960).

18. R. Kh. Burshtein and L. A. Larin, Zhur. Fiz. Khim. <u>32</u>, 194 (1958).

19. J. Bardeen, Phys. Rev. <u>71</u>, 717 (1947).

20. J. A. Dillon, Bull. Am. Phys. Soc. <u>1</u>, 53 (1958).

21. J. A. Dillon and H. E. Farnsworth, J. Appl. Phys. <u>28</u>, 174 (1957).

22. N. F. Mott, Trans. Faraday Soc. <u>43</u>, 422 (1947).

23. W. Brattain and J. Bardeen, Bell System Tech. J. <u>32</u>, 1 (1953).

24. H. H. Madden and H. E. Farnsworth, Phys. Rev. <u>112</u>, 793 (1958).

25. A. F. Gibson, P. Aigrain, and R. E. Burgess, Progress in Semiconductors <u>1</u>, 165 (1956).

26. T. I. Galkina, Fiz. Tverd. Tela <u>1</u>, 216 (1959).

27. C. G. B. Garrett and W. H. Brattain, Bell System Tech. J. <u>35</u>, 1041 (1956).

28. N. A. Shurmovskaya and R. Kh. Burshtein, Doklady Akad. Nauk SSSR <u>129</u>, 172 (1959).

29. R. Kh. Burshtein, Electrochim. Acta 1962 (in press).

KINETICS OF OXYGEN CHEMISORPTION ON SEMICONDUCTORS

É. Kh. Enikeev

Chemical Physics Institute, Academy of Sciences, USSR

It may be regarded as established that the rate of chemisorption decreases with increase of surface coverage θ much faster than is predicted by Langmuir's kinetics. Usually the rate of gas chemisorption on semiconductors is described by the logarithmic Roginskii—Zel'dovich equation

$$q = \frac{2.3}{\alpha} \log (t + t_0) - \frac{2.3}{\alpha} \log t, \tag{1}$$

where $t_0 = 1/a\,\alpha$, and a and α are constants, or by the power equation of Bangham

$$q = At^{1/n}, \tag{2}$$

where $n = 2$. These phenomenological equations are valid for widely differing adsorption mechanisms, provided the barrier to chemisorption rises linearly with surface coverage in (1) and logarithmically in (2) [1].

Consider, for example, the following: a surface nonuniformity with centers distributed according to their activation energies, the distribution being characteristic for a given adsorbent; or the existence of an interaction of the adsorbed atoms with one another through the electron gas of the crystal. Such mechanisms allow us to describe the chemisorption kinetics equally well by either of the above two equations.

Isotopic methods have indicated the considerable importance of the nonuniformity in the kinetics of chemisorption [2].

In several calculations the effect of charging of the surface during adsorption on the kinetics of the process has been pointed out [3, 4]. However, the present author is not aware of any experimental work giving a direct comparison between the variation of the chemisorption rate with increase of the surface coverage and the variation of the height of the chemisorption potential barrier.

Investigations of the long-lived relaxation of the surface potential of a semiconductor, disturbed from its equilibrium state by illumination, application of an external electric field or a sudden change of temperature in the presence of active gases, indicate that other explanations are possible, since under these conditions the degree of surface coverage by adsorbed gases was not altered.

The present author investigated the kinetics of oxygen chemisorption on MnO_2, and in lesser detail on ZnO and NiO. Experiments were carried out as follows. A powder sample was placed in a special cell and was evacuated at high temperature to $\sim 10^{-6}$ mm Hg. On the surface of an oxide semiconductor treated in this way there remains, apart from other residual gases, a certain amount of oxygen θ_0. The value of θ_0 for a given semiconductor depends on the conditions of outgassing (temperature, duration of evacuation, final vacuum, etc.). Consequently it was necessary to use different outgassing conditions for different oxide semiconductors and at the same time to enforce rigid standardization of the conditions in order to obtain sufficiently reproducible results.

MnO$_2$ was evacuated at 230°C, since above this temperature it is transformed into Mn$_2$O$_3$ [5]. ZnO and NiO were usually outgassed at 400°C.

At these temperatures (< 0.5T$_{m.pt.}$) the oxygen was removed mainly from the surface layers [6], but this was sufficient to obtain reproducible results.

After being outgassed and cooled to room temperature the sample was placed by means of an electromagnet under a vibrating electrode, so that its contact potential difference (c.p.d.) could be measured. Oxygen was then admitted to the cell and from this moment the variation of the c.p.d. and the reduction of the O$_2$ pressure in the system were recorded continuously.

The initial pressure of oxygen, $p^0_{O_2}$, was 1 mm Hg and the reduction of the pressure at the end of the experiment, Δp, did not exceed 10-15% $p^0_{O_2}$. The duration of the experiment was not less than two hours.

The variation of the c.p.d. represented the variation of the contact potential φ of the test surface provided the values of φ of the reference electrode did not vary in the atmosphere of the test gas.

Several workers have suggested coating the reference electrode with some inert substance (e.g., calcium palmitate [7] or glass [8]) which does not adsorb certain gases.

On the other hand it has been suggested that oxygen is weakly adsorbed on gold [9]. Lyashenko [10] measured the value of φ of a gold reference electrode using an electron gun and found that O$_2$ increases φ_{Au} by 30-50 mV. To estimate the variation of φ of a gold electrode the present author carried out additional experiments in the following way.

A gold plate, cut from the same material as the reference electrode, was placed on the lower (nonvibrating) electrode.

The initial c.p.d. between the electrode and the plate was usually less than 0.1 V. Then the support with the gold test plate was placed in the heated part of the cell and the plate was oxidized (at ≈1 mm Hg of oxygen at various temperatures).

To measure the c.p.d. the support with the test plate was moved for a short time to the cold part of the cell, and measurements were carried out there; it was then again subjected to high-temperature treatment in oxygen. Figure 1 shows several $\Delta V_C = f(t)$ curves obtained in this way. Curve 1 in Fig. 1 represents the variation of φ of a plate which was initially in air and was then heated at 100°C in oxygen at 1 mm Hg pressure. The value of φ rose by 0.2 V during the first 3-4 min and during the next 210 min it rose by a further 0.18 V.

Cooling of the plate to room temperature and evacuation did not alter the Au/Au$_{oxid.}$ contact potential difference. Next the outgassing temperature was raised to 250°C and the c.p.d. returned to its initial value (0.05 V) in 2 hours (this is not shown in Fig. 1). Measurements were carried out in the same way as in the oxidation experiments, i.e., with a cold reference electrode.

At higher temperatures (300°C) the oxidation process was considerably faster and was practically complete after 30-40 min (curve 2). Outgassing at this temperature did not restore the original c.p.d. value (curve 3): φ of the oxidized gold fell by only 70-80 mV. Increase of the oxidation temperature to 360°C did not alter the c.p.d. (0.5 V) during the first 15 min (this is not shown in Fig. 1). With increase of the oxygen pressure to 2.5 mm Hg the value of $\Delta\varphi$ increased to 0.55 V and after the next 10 min it rose to 0.58 V. After increase of the temperature to 400°C the value of $\Delta\varphi$ in 1 mm Hg of oxygen rose to 0.90 V after 2 hours (curve 4).

After reaching the latter value, 0.90 V, the c.p.d. remained practically constant: heating in vacuum at 250°C for one hour reduced φ by at most 0.06 V (curve 5), while heating in an atmosphere of oxygen at the same temperature raised φ by 0.06 V (curve 6); evacuation of oxygen at room temperature did not alter φ of the oxidized gold (curve 7).

The oxide film was partly reduced by heating for 5 hours at 160°C in an atmosphere of H$_2$. The new value of the c.p.d. (~0.4 V) did not change in the next 5 hours of heating in H$_2$. Heating of the oxidized gold in CO altered the c.p.d. by only 0.2 V. Thus our results indicate that after preliminary oxidation of the gold electrode at ~400°C the variations in φ_{Au} do not exceed ~0.1 V on heating either in vacuum or in oxygen ($p_{O_2} \approx 1$ mm) up to 250°C.

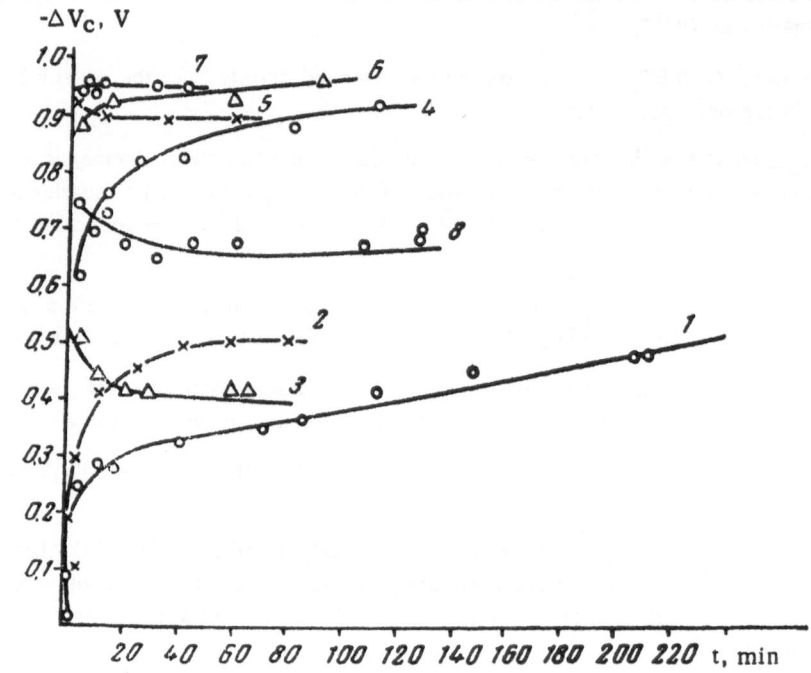

Fig. 1. Variation of the contact potential difference during high-temperature treatment of one of two gold plates in oxygen and in vacuum.

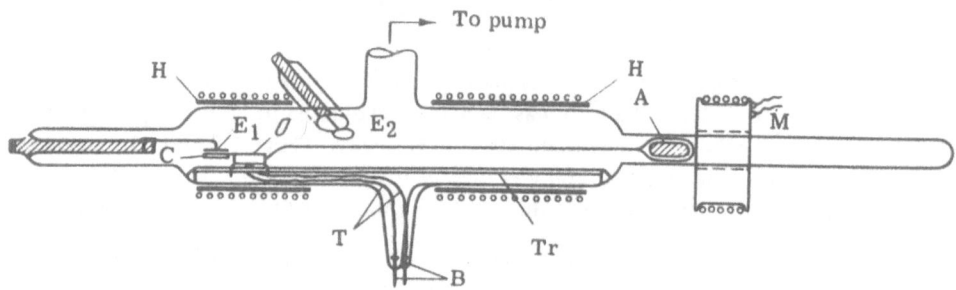

Fig. 2. Cell for measuring contact potential difference. E_1 and E_2 are the heated and cold gold electrodes; C is the glass-coated molybdenum plate; A is the iron core; M is the electromagnet; T is the thermocouple; H is the heater; Tr are the tracks along which the sample 0 is moved.

The results obtained were checked additionally with a cell specially constructed for this purpose (Fig. 2).

In contrast to the previous cell, this new cell has an additional unheated gold electrode E_2 and a molybdenum electrode C protected from the effect of the ambient atmosphere by a thin layer of molybdenum glass. The cell so constructed, and measurements of the c.p.d. between the gold electrodes E_1 and E_2, one of which is heated in an oxygen atmosphere and the other of which is cold, as in the earlier tests, allowed us to compare the contact potentials of the two gold electrodes with respect to the glass-coated electrode.

Curve 8 in Fig. 1 shows the variation of the c.p.d. between a gold electrode oxidized in an atmosphere of O_2 (1 mm Hg) and a glass-coated electrode C at 250°C. * As in the case of curve 6, an increase of φ_{Au} by 0.07-0.08 V was observed.

---

* The signs of ΔV_C are different since in the first case we heated the plate placed on the support, while in the second we heated the vibrating plate.

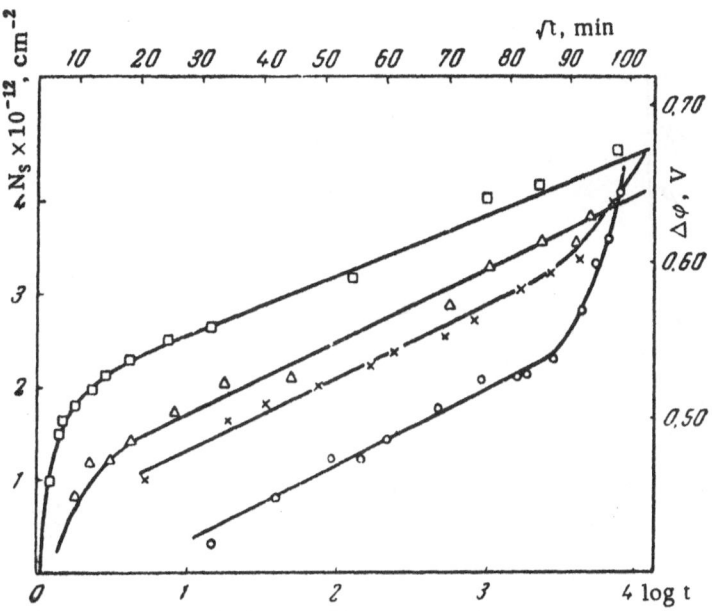

Fig. 3. Kinetics of oxygen adsorption in manganese dioxide at 100°C and variation of the work function. From the top downwards the curves represent: 1) $\Delta \varphi \approx f(\sqrt{t})$; 2) $N_s \approx f(t)$; 3) $\Delta \varphi \approx f(\log t)$; 4) $N_s \approx f(\log t)$.

The agreement between the results obtained by the two different methods indicates their sufficient reliability, and allows us to determine the variation of the electron work function of a semiconductor from measurements of the c.p.d. in an oxygen atmosphere.

Figure 3 shows the results on the kinetics of oxygen chemisorption on MnO_2 at 100°C. It is clear from the figure that initially the amount of oxygen adsorbed (N_s) rises linearly with log t, but at the end of the experiment the curve deviates considerably from linearity.

The value of $\Delta \varphi$ varies in the same way with time. The ends of the curves in Fig. 3 become linear in coordinates of N_s or $\Delta \varphi$ versus $\sqrt{t}$. With increase of temperature the inflection point is displaced toward shorter times and finally at 200°C practically all the experimental points fit a linear dependence.

Figure 4 shows the dependences $N_s = f(\sqrt{t})$ for various temperatures. This figure indicates that the rate of the process increases considerably with increasing temperature.

The same dependence of $\Delta \varphi$ on time is observed on heating MnO_2 at a temperature higher than T, the temperature of the outgassing treatment (Fig. 5).

Figure 6 gives the change in the work function of MnO_2 on admission of successive quantities of oxygen into the cell at 200°C. During the first minutes after the admission of oxygen φMnO_2 rose strongly. After complete absorption of the gas admitted into the cell (the pressure did not exceed $1-2 \times 10^{-4}$ mm Hg) φ began to decrease but even after 40 min the steady-state value of φ was not reached. During the admission of the subsequent portions of oxygen the reduction of φ after complete absorption became smaller and smaller, and finally, after the fourth portion of oxygen no reduction of φ was observed during the next 15 min.

All these experimental observations (Figs. 3-6) indicate that volume absorption of oxygen in manganese dioxide begins to play an important role above 100°C.* The direction of the process of oxygen diffusion (into

---

*In this case it would be more correct to express the amount of absorbed gas not in units of N_s (molecules/ cm²) but in $N_s S/V$, where S is the total area in cm² and V is the volume of the sample in cm³. However we retained the units of N_s for convenience in subsequent comparisons with the adsorption measurements.

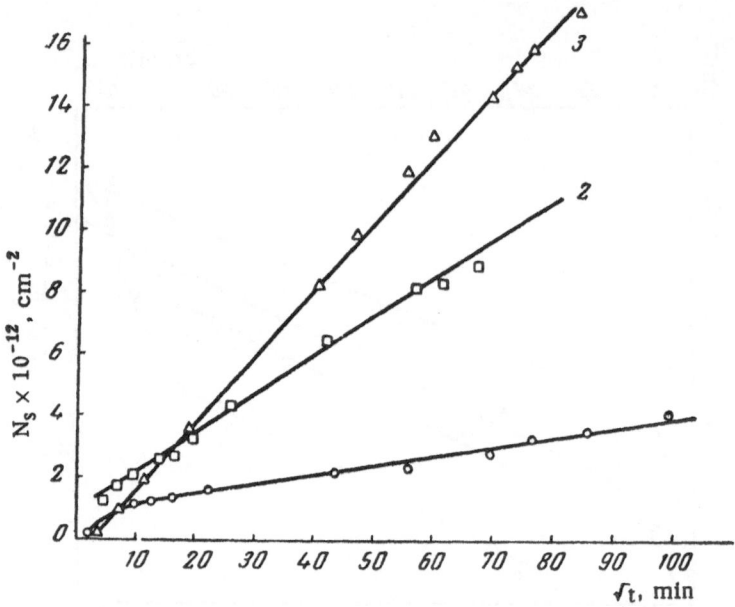

Fig. 4. Kinetics of oxygen absorption at various temperatures (in °C):
1) 100; 2) 180; 3) 215.

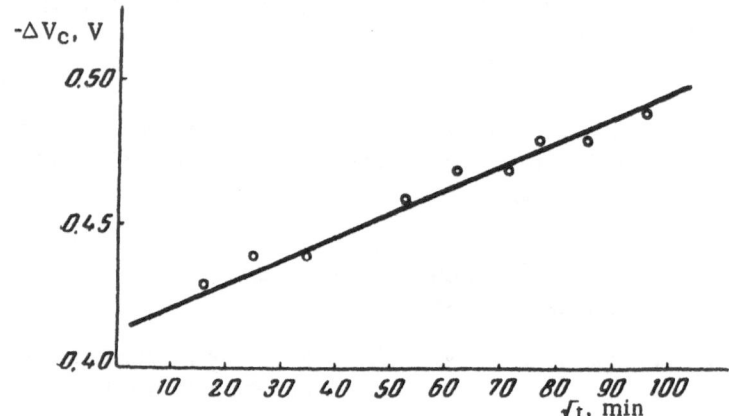

Fig. 5. Variation of the contact potential ΔV_C with the duration of
heating of manganese dioxide in vacuum at 230°C.

grain interiors or evolution as the gas phase) depends on the ratio of the surface and volume concentrations of O_2 in the grain.

The calculated activation energy for diffusion, E_g, is 9 kcal/mole, which is equal to the value found earlier [11].

The dissolving of oxygen in MnO_2, which is considerable above 100°C, restricts the range of temperatures available for investigating the intrinsic kinetics of oxygen chemisorption on MnO_2. Figures 7-9 give the results for the chemisorption of O_2 at 20, 55 and 90°C, respectively. It is clear that at these temperatures there are no inflections in the kinetic curves and N_s, $\Delta \varphi$ vary linearly with log t. Eliminating time from the kinetic curves $N_s = f_1(\log t)$ and $\Delta \varphi = f_2(\log t)$, we find a linear dependence of $\Delta \varphi$ on N_s which is identical with the dependence of $\Delta \varphi$ on N_s obtained from the steady-state values of $\Delta \varphi$ and N_s (Fig. 10, the flat parts of the curves).

* Surface coverages of up to ~1×10^{11} cm^{-2}, which correspond to the steep part of the $\Delta \varphi = f(N_s)$ curve in Fig. 10, were reached instantaneously, and we were unable to measure the adsorption kinetics during this stage.

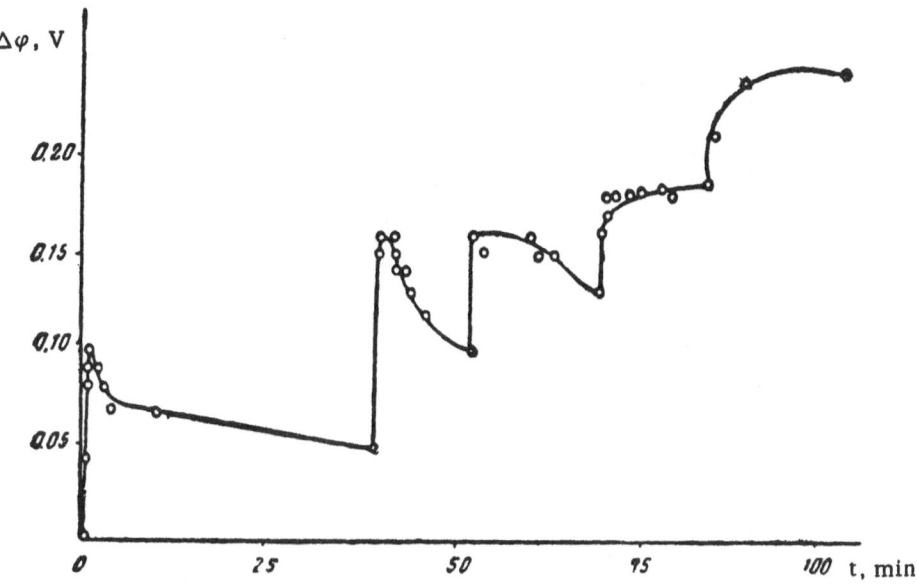

Fig. 6. Variation of the work function $\Delta\varphi$ of manganese dioxide with time from the admission of successive portions of oxygen into the cell at 200°C.

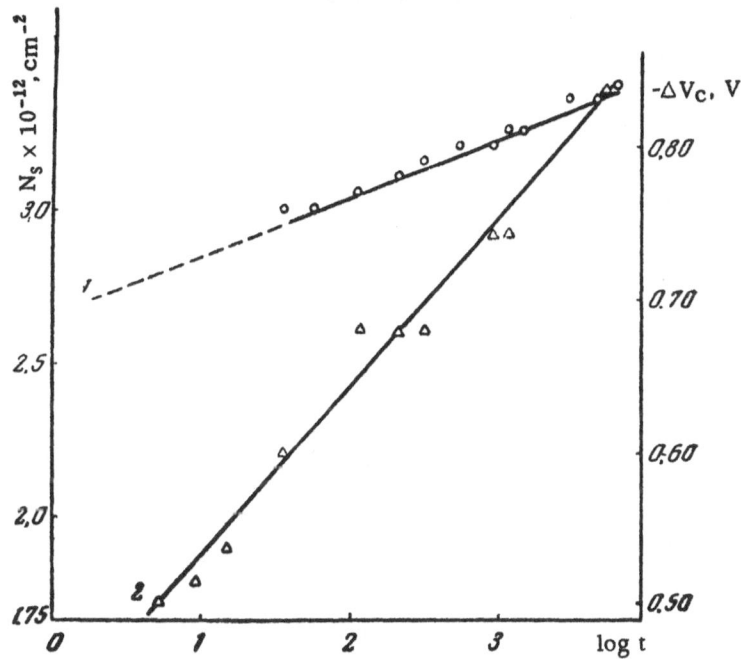

Fig. 7. Kinetics of oxygen sorption on manganese dioxide (1) and variation with time of the contact potential of manganese dioxide during adsorption of oxygen (2) at 20°C.

Thus we find that the curves of oxygen sorption by manganese dioxide have three distinct stages. The first represents the practically instantaneous adsorption which takes place without an activation energy. The surface coverages are very low (up to $10^{11}\,\text{cm}^{-2}$) and φ rises rapidly by 200-250 mV. During the second stage the chemisorption process is extended in time and is governed by a logarithmic isotherm. The work function increases linearly with the surface coverage but $d\varphi/dN_s$ is smaller than in the first stage, while N_s reaches

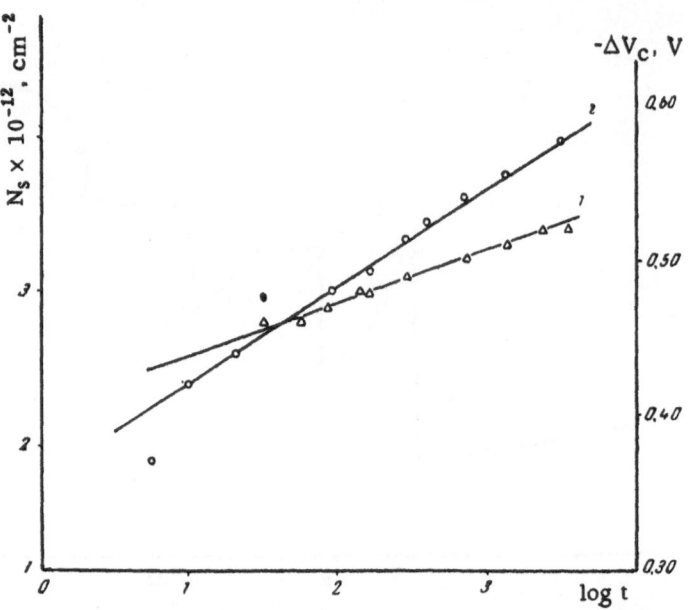

Fig. 8. Kinetics of oxygen adsorption on manganese dioxide
(1) and variation of the contact potential with time during
adsorption of oxygen (2) at 55°C.

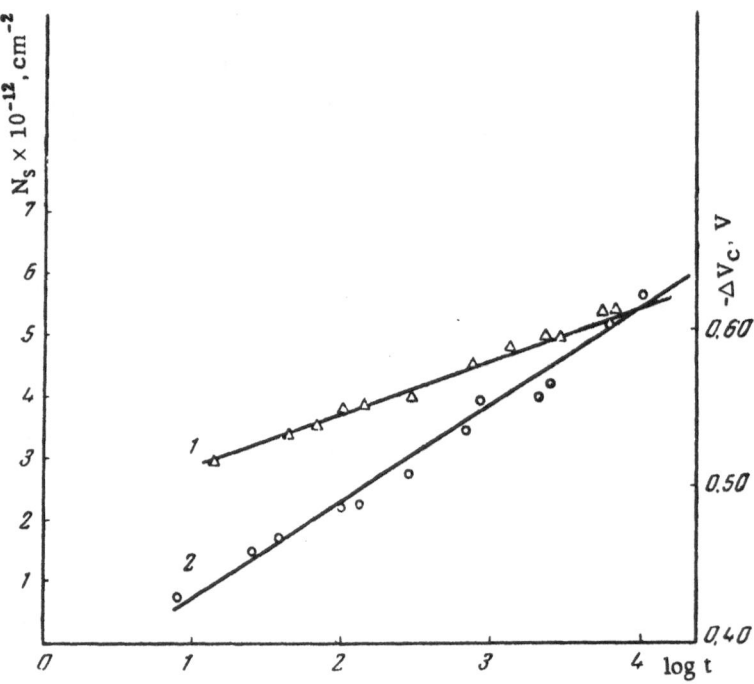

Fig. 9. Kinetics of oxygen adsorption on manganese dioxide (1) and
variation of the contact potential with time during adsorption of oxygen (2) at 90°C.

$3\text{-}6 \times 10^{12}\,\text{cm}^{-2}$. Beginning from 100°C the third, and slowest, mechanism becomes active; it represents the dissolving of oxygen in the MnO_2 lattice (during this stage φ also increases but only slightly).

The logarithmic dependence of the amount of adsorbed gas on time, given by Eq. (1) was found experimentally as long ago as 1934 by Zel'dovich and Roginskii [12] for the adsorption of CO on MnO_2; later this dependence has been found to be more generally valid in chemisorption kinetics. The kinetics of oxygen adsorption on MnO_2 has also been found to obey the logarithmic equation (1) [12].

However, Bruns and co-workers [11] showed that the rate of chemisorption of CO is governed by the diffusion of oxygen from the grain interiors to the surface followed by the oxidation of CO to CO_2. The high mobility of oxygen in the MnO_2 lattice, and the concentration gradient produced during the reduction of the MnO_2 surface by carbon monoxide, ensured a high rate of adsorption (oxidation) of CO. The characteristic phenomenon of "relaxation," consisting of the regeneration of the adsorption properties of the surface due to the inflow of oxygen from the grain interiors, was observed by Bruns et al. and also in present work; we found it as an increase of φ of an outgassed sample on heating in vacuum (Fig. 5).

The dependence of $\Delta\varphi$ on $\sqrt{t}$ confirms the diffusion mechanism of the "relaxation" phenomenon. However, at relatively low temperatures (100°C) the value of φ of outgassed MnO_2 rises only a little (30-50 mV/hr). Obviously the volume diffusion of oxygen at these temperatures is slow and therefore the absorption of oxygen from the gas phase is governed mainly by the adsorption of the gas, which we shall now consider (Figs. 7, 8, and 9).

Considering the chemisorption of oxygen on a semiconductor as the reaction

$$O_2^{gas} + e^- \rightleftarrows O_2^-\text{ads}$$

we shall write the expression for the chemisorption rate in accordance with the law of mass action:

$$\frac{dN_s}{dt} = \frac{p}{\sqrt{2\pi MkT}} \tau C_n \overline{V} \left(1 - \frac{N_s}{N_{tot}}\right) n_s(N_s) - BN_s. \tag{3}$$

Here $\dfrac{p}{\sqrt{2\pi MkT}}$ is the number of collisions of O_2 molecules per second with 1 cm^2 of the surface; $[1 - (N_s/N_{tot})]$ is the fraction of the surface which is free; $C_n\overline{V}$ is the capture coefficient, where C_n is the cross section for electron capture and $\overline{V}$ is the thermal velocity of electrons; BN_s is a term which allows for oxygen desorption; n_s is the density of equilibrium electrons at the surface; $n_s(N_s)$ is a function, as yet unknown, of the surface coverage N_s; τ is the mean lifetime of a molecule on the surface, governed by electrostatic or polarization forces.

For conditions remote from equilibrium and at not too high surface coverages, which is the case in "depleting" adsorption, we can rewrite Eq. (3) as follows:

$$\frac{dN_s}{dt} = C \exp\left\{-\frac{eV_d(N_s)}{kT}\right\}, \tag{4}$$

where V_d is the energy-band curvature at the semiconductor surface;

$$C = \frac{p}{\sqrt{2\pi MkT}} C_n\overline{V} N_{eff} \exp\left(\frac{E_F - E_C}{kT}\right).$$

Neglecting the contribution of the induced dipole moment to $\Delta\varphi$, we have

$$\Delta\varphi = eV_d = \gamma N_s, \quad \text{where} \quad \gamma = \frac{d\varphi}{dN_s}.$$

Introducing $\alpha = \gamma/kT$, we find that

$$\frac{dN_s}{dt} = C \exp\left(-\alpha N_s\right). \tag{5}$$

Equation (5) is identical with the kinetic isotherm of Roginskii and Zel'dovich, and the coefficients in that equation have definite physical meaning and may be checked experimentally. For example, the coefficient α, representing the change of potential per unit surface charge, is the reciprocal of the combined capacitance of the space-charge layer and the surface states.

We can write in general:

$$\frac{1}{C} = \frac{1}{e}\frac{d\varphi}{dN_s} = \frac{1}{C_{sc} + C_{ss}}, \tag{6}$$

where C_{ss} is the capacitance of the surface states, and C_{sc} is the capacitance of the space-charge layer.

If $C_{ss} < C_{sc}$, then α is governed by the concentration of defects in the space-charge layer. This case is of special interest for impurity semiconductors.

We note that charging of the surface governs the deviation of the chemisorption kinetics from the relations given by Langmuir if, in general, $\frac{d\varphi}{dN_s}N_s > kT$. In this case $\exp\left\{-\frac{eV_d(N_s)}{kT}\right\}$ decreases faster than $[1 - (N_s/N_{tot})]$ in Eq. (4).

Differentiating dN_s/dt logarithmically with respect to temperature we can find the chemisorption activation energy, E:

$$E = E_0 + \gamma N_s. \tag{7}$$

Thus the activation energy consists of two parts: a constant E_0, which is the intrinsic activation energy, and a variable part increasing with the surface coverage N_s.

From Eq. (7) it follows that $dE/dN_s = \gamma > 0$. Such a relationship is typical of collective processes. In other words the interaction of adsorbed atoms (molecules) with one another through the electron gas of the crystal reduces the rate of adsorption.

The question now is whether this is the governing factor in the reduction of the chemisorption rate in our case.

To answer this question we measured the temperature dependence of the chemisorption rate (Figs. 7, 8, and 9) and, using the equation $\ln\frac{t_2}{t_1} = \frac{E}{R}\left(\frac{1}{T_2} - \frac{1}{T_1}\right)$, calculated the activation energies. On increase of the surface coverage, for example from $N_s = 2 \times 10^{12}$ to $N_s = 4 \times 10^{12}$ cm^{-2}, the activation energy for chemisorption increased by 2.5 kcal/mole, while in the same range of surface coverage $\gamma N_s \approx 0.1$ eV = 2.3 kcal/mole. The agreement between these two values is obviously very good if we remember that the range of temperatures and of ΔE is small.

We shall write the chemisorption rate equation in the following form:

$$N_s = \frac{2.3}{\alpha}\log t \tag{8}$$

Because

$$\alpha = \frac{d\varphi}{dN_s}\cdot\frac{1}{kT},$$

we have

$$\frac{\Delta\varphi}{\log t} = 2.3\,kT.$$

Figure 11 gives the experimental results in coordinates of $\Delta\varphi/2.3\log t$ and T°K.

The slope of the straight line drawn through the experimental points in Fig. 11 is 1×10^{-4}, which is slightly greater than Boltzmann's constant k (k = 0.87×10^{-4} eV/deg).

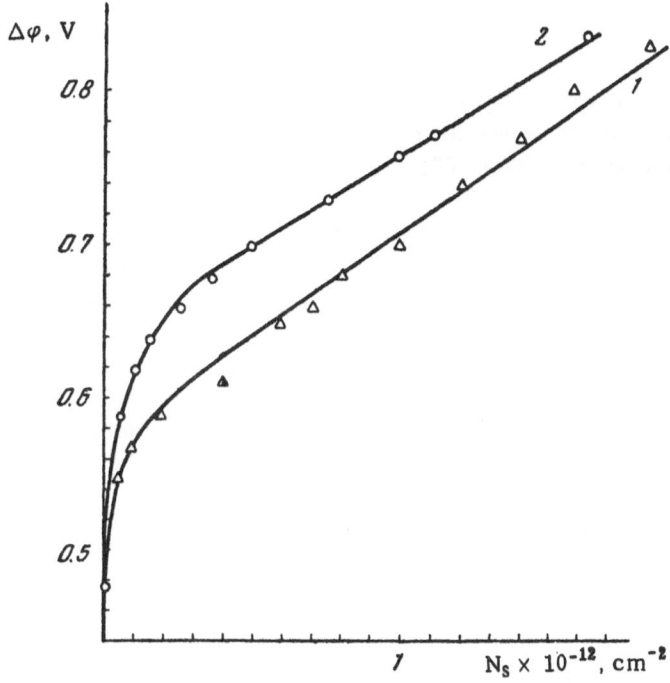

Fig. 10. Dependence of the change in the work function of manganese dioxide as a function of the surface coverage by oxygen at 20°C (1) and at 90°C (2).

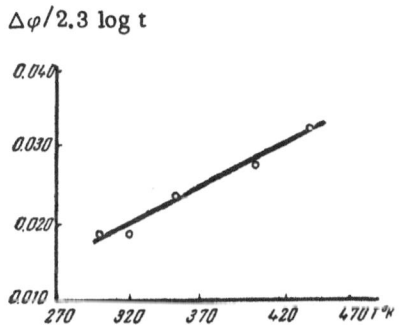

Fig. 11. Dependence of the quantity $\Delta\varphi/2.3 \log t$ on temperature for t = 100 sec.

The experimental results on the kinetics of oxygen chemisorption on MnO_2 reported in the present paper allow us to conclude that the rate and nature of chemisorption are governed by the charging of the semiconductor surface.

We also measured the rate of oxygen chemisorption on nickel oxide at 300°C. We found that the kinetics of this process is governed by the empirical equation of Bangham: $N_s = At^{1/n}$. As established earlier [13], in this case $\Delta\varphi \approx \beta \log(N_s)$ at 20°C. Substituting this expression into Eq. (4), we obtain

$$\frac{dN_s}{dt} = C \exp\left\{-\frac{\beta \lg N_s}{kT}\right\}.$$

Assuming $\beta = kT$ [14], we can easily show that $N_s = A't^{1/n}$, where n = 2 in agreement with the Bangham isotherm. Finally, for ZnO, the linear dependence of φ_{ZnO} on $N_{s(O_2)}$ obtained by the present author [13] agrees with the kinetic Roginskii—Zel'dovich isotherm for the adsorption of oxygen on ZnO [15, 16].

It should be stressed that a detailed study (such as was done for MnO_2) of the adsorption of oxygen on NiO and ZnO, which is reasonably fast only at high temperatures, is not very useful if there is no direct control of the changes of φ of the reference electrode. However, it is precisely such experiments which are of great interest because of the possible considerable variations of E and T.

LITERATURE CITED

1. S. Z. Roginskii, Adsorption and Catalysis on Inhomogeneous Surfaces (Izd. AN SSSR, 1948).
2. N. P. Keier and S. Z. Roginskii, Izvest. Akad. Nauk SSSR, Otdel. Khim. Nauk 27 (1950).
3. P. B. Weisz, J. Chem. Phys. 21, 1531 (1953).
4. K. Hauffe, Advances in Catalysis 7, 213 (1955).
5. W. E. Garner and T. Ward, J. Chem. Soc. 857 (1939).
6. D. I. M. Bevan, I. P. Shelten, and I. S. Anderson, J. Chem. Soc. 1729 (1948).
7. K. Lion, Instrument in Scientific Research. Electrical Input Transducers (New York, 1959), p. 215.
8. R. Kh. Burshtein and L. Larin, Zhur. Fiz. Khim. 32, 194 (1958).
9. B. M. W. Trapnell, Proc. Roy. Soc. (London) A218, 566 (1953).
10. V. I. Lyashenko, Dissertation (Kiev, 1955).
11. N. A. Shurmovskaya, B. P. Bruns, and Z. Ya. Mel'nikova, Zhur. Fiz. Khim. 25, 1306 (1951).
12. S. Roginskii and I. Zel'dovich, Acta Physicochim. 1, 554 (1934).
13. É. Kh. Enikeev, Problemy Kinetiki i Kataliza 10, 88 (1960).
14. V. B. Sandomirskii, Dissertation (M., 1958).
15. É. Kh. Enikeev, L. Ya. Margolis, and S. Z. Roginskii, Doklady Akad. Nauk SSSR 129, 372 (1959).
16. S. R. Morrison, Advances in Catalysis 7, 259 (1955).

EFFECT OF ADSORPTION OF POLAR
MOLECULES ON THE SURFACE PROPERTIES
OF GERMANIUM

Yu. F. Novototskii-Vlasov and
M. P. Sinyukov

P. N. Lebedev Physics Institute, Academy of Sciences, USSR

The desorption of water increases the concentration of surface recombination centers in germanium. This was found in experiments on the vacuum heating of germanium samples [1-4]. After heating at 500°K the maximum surface recombination velocity (s_{max}) increases by a factor of 10-15 compared with the initial value, but the form of the curve $s(Y_s)$, where Y_s is the electrostatic surface potential, is not altered. This indicates an increase by a factor of 10-15 of the concentration of surface recombination centers without any change occurring in their nature.

If a sample, previously heated at 500°K, is placed in an atmosphere of water vapor, then, after a certain time, the surface recombination velocity decreases so that eventually the $s(Y_s)$ curve practically retraces the initial curve before heating. Thus water vapor is shown to be capable of neutralizing recombination centers, leaving only 5-10% of their maximum number still active.

The reversibility of the surface properties during the heating–water-vapor cycle suggests that the neutralization of recombination centers is due to water physically adsorbed on the surface of germanium. Since water is a polar liquid, in the case of physical adsorption the electric field of a water molecule which has approached sufficiently close to a recombination center may alter the parameters of this center so as to destroy its recombination activity. However, we cannot exclude the possibility of chemical interactions since water, on the one hand, dissolves the oxide layer, and, on the other, may be bound to the surface forming germanium hydroxide. In the latter case the radical OH may play the decisive role.

In order to find which neutralization mechanism is active, it is necessary to replace water molecules on the germanium surface with molecules of other substances possessing some of the properties of water. The replacement of the adsorbed water on the germanium surface with alcohol molecules avoids any possibility of dissolving the oxide layer on the germanium. By using polar substances which do not contain the radical OH we can avoid both "chemical" interactions leaving open only the possibility of an interaction between the molecular electric dipole and the recombination center.

Experimental Technique

When placing germanium samples in atmospheres of various vapors, special attention was paid to the absence of traces of water in these vapors. For this reason we selected liquids having lower room-temperature vapor pressures than the vapor pressure of water. Then, even if water is accidentally admitted into a cell with

TABLE 1

Liquid	Chemical formula	Dipole moment 10^{18} abs. esu	Vapor pressure at 300°K, mm Hg
N-amyl alcohol	$C_5H_{11}OH$	1.6	3
Isoamyl alcohol	$(CH_3)_2CH-CH_2-CH_2OH$	1.8	4
Chlorobenzene	C_6H_5Cl	1.6	15
Nitrobenzene	$C_6H_5NO_2$	3.9	$3 \cdot 10^{-1}$
Ethyl ether	$(C_2H_5)_2O$	1.2	550
Benzene	C_6H_6	0	100
Water	H_2O	1.8	30

a purified liquid, the water can easily be removed by pumping out the cell. The amounts of water and other liquids in the vapor phase are proportional to their vapor pressures, so that prolonged pumping out of the cell containing the liquid removes the water completely from the pumped volume.

On the basis of these considerations, we selected the following liquids: normal amyl alcohol, isoamyl alcohol, chlorobenzene, and nitrobenzene. Some of their properties are listed in Table 1. The same table contains the data for water, ethyl ether, and benzene.

Benzene was used in a control experiment to determine the efficiency of the method adopted for removing traces of water from liquids by vacuum pumping. Table 1 shows that the conditions for the removal of traces of water from benzene are less favorable than for the first four liquids listed in the table since the vapor pressure of benzene at room temperature is higher than the vapor pressure of water.

The conditions for the removal of water by vacuum pumping were least favorable in the case of ethyl ether. Special attention was therefore paid to the careful preliminary purification of the ethyl ether and drying of the cell immediately before filling it with ether. The preliminary purification of ethyl ether by repeated distillation and selection of the middle fractions resulted in a liquid free from water. After placing this pure ether into a dried working cell we evacuated the cell repeatedly in order to remove possible water traces absorbed by the ether from the air. Since the vapor pressure of ether is considerably greater than the vapor pressure of water, this procedure produced evaporation of the ether itself. However, the small amount of ether remaining after vacuum purification was sufficient for our experiments.

The experiments were carried out on samples of p-type germanium of resistivity 28-30 $\Omega \cdot$cm and having a volume carrier lifetime of 500-700 μsec. The samples were etched in H_2O_2 with an admixture of NaOH, washed in deionized water (resistivity 6-7 M$\Omega \cdot$cm) and dried with alcohol and ether. After being exposed to atmospheric air for several hours, in order to stabilize the surface, each sample was clamped in a holder, placed in a working cell, which was evacuated to about 10^{-6} mm Hg. Under these conditions we recorded the dependence of the surface recombination velocity, and of the charge captured by fast surface states, on the electrostatic surface potential Y_s, i.e., we recorded $s(Y_s)$ and $Q_{ss}(Y_s)$.

The samples were next heated to 500°K, kept at this temperature for 5 min and cooled to 300°K, after which the curves $s(Y_s)$ and $Q_{ss}(Y_s)$ were recorded again. The working sample cell was then connected to another cell containing the test liquid and we recorded the changes, due to the adsorption of the test-liquid molecules on the sample surface, in the surface recombination velocity and the charge captured by the fast surface states.

All measurements were carried out at 300°K by the usual method of combining the field effect for strong sinusoidal signals with the steady-state photoconductivity [5].

Experimental Results

The experiments on the amyl alcohols showed that the vapors of normal and isoamyl alcohols behave similarly to water vapor and practically completely neutralize the recombination centers produced by heating the sample at 500°K. After exposure of the sample to an atmosphere of amyl alcohol vapor for one hour the $s(Y_s)$ curve almost completely coincided with the initial $s(Y_s)$ curve recorded before heating.

As mentioned above, these experiments did not completely exclude the possibility of chemical neutralization, since the OH radical, which occurs in alcohol molecules, could be responsible for the effects observed. The main emphasis was therefore placed on a study of the effect on surface recombination of chlorobenzene and nitrobenzene vapors since the chemical structure of these liquids has nothing in common with the structure of water. The results of the experiments using these liquids are given in Fig. 1.

Figure 1 shows that chlorobenzene completely neutralizes the recombination centers introduced by heating. On the other hand, although s_{max} after treatment with nitrobenzene vapor was considerably smaller than s_{max} after heating at 500°K, it was still about twice as large as the initial value of s_{max}. This result may be explained by the fact that, as indicated in Table 1, the vapor pressure of nitrobenzene at room temperature is considerably smaller than the vapor pressures of other liquids, though the duration of treatment of the samples in various vapors (one hour) was the same for all liquids.

We therefore carried out additional experiments on nitrobenzene vapor, in the course of which we recorded the time variations of the surface parameters. It should be noted that nitrobenzene is unusual, compared with the other polar liquids investigated here and compared with water, in its band shift at the surface. It is known that water vapor induces negative charge on the surface and shifts the surface bands in the direction of stronger n-type conduction. As a result of this it is very difficult to reach the minimum surface conductivity and carry out quantitative measurements in humid atmospheres. Moreover, the maximum of the surface recombination velocity lies usually near the minimum of the surface conductivity. All these circumstances make it impossible to investigate the kinetics of the neutralization of surface recombination centers in humid atmospheres.

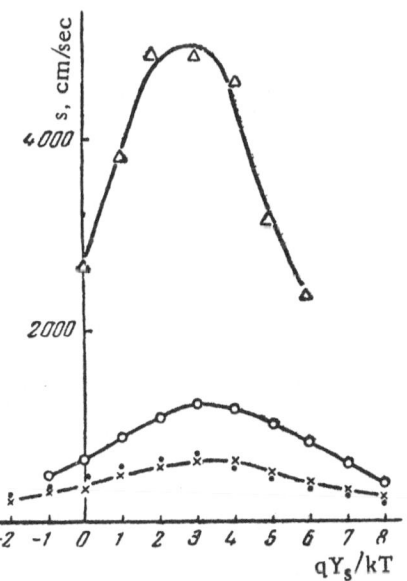

Fig. 1. Changes in the surface recombination velocity on exposure to atmospheres of chlorobenzene and nitrobenzene vapors: ● - in vacuum before heating; △ - after heating at 500°K; × - in chlorobenzene; ○ - in nitrobenzene.

Nitrobenzene, on the other hand, charges the surface positively. The surface potential of a sample in nitrobenzene vapor is close to the surface potential in vacuum. This made it possible to investigate the kinetics of the neutralization of recombination centers by nitrobenzene molecules. Moreover, owing to the low vapor pressure of nitrobenzene, the process of neutralization proceeds relatively slowly, which avoids the possibility of different changes occurring in different regions of the sample surface.

Results on the effect of treatment with nitrobenzene vapor on the surface recombination velocity of a sample which had earlier been heated at 500°K are shown in Fig. 2. After 20 hours the recombination centers introduced by heating are practically completely neutralized. The form of the $s(Y_s)$ curve is then similar to the initial curve.

The experimentally observed time dependence of the maximum surface recombination velocity can be represented by a simple power law

$$s_{max} = a + bt^{-1/3},$$

where a and b are constants. Figure 3 gives the experimental points for s_{max} of nitrobenzene plotted against $t^{1/3}$. They fit very well a straight line which passes through the origin of coordinates. This means that, as $t \to \infty$, all the centers should be neutralized. However, we

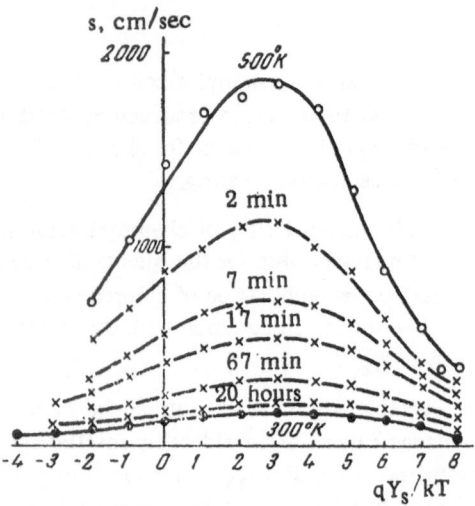

Fig. 2. Changes in the surface recombination velocity on adsorption of nitrobenzene vapor on the sample surface. Numbers by the curves indicate the duration of exposure to nitrobenzene vapor.

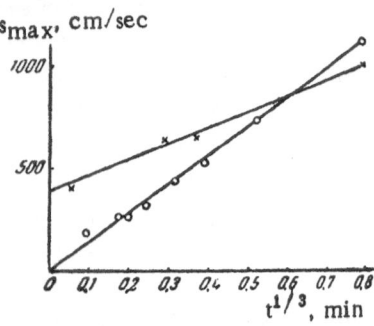

Fig. 3. Dependence of the maximum surface recombination velocity on time: O - nitrobenzene; × - ethyl ether.

see that the point corresponding to t = 20 hours is a considerable distance from the straight line. Obviously nitrobenzene molecules cannot reach some of the centers.

There is an interesting relationship between the surface recombination velocity and the charge captured by the fast surface states (Figs. 2 and 4). Two minutes after the admission of nitrobenzene vapor the surface recombination velocity had fallen by a factor of about 2, while the charge captured by the fast states remained the same. A measurement carried out 7 min after the admission of nitrobenzene vapor showed a further reduction of the surface recombination velocity and a strong fall of the charge captured by the fast surface states. It seems that after 2 min from the admission of nitrobenzene vapor some of the centers cease to participate in recombination but continue to act as capture centers. Later these centers lose also their ability to capture charge.

The results of the experiments with ether are similar to the results for other polar liquids. The value of s_{max} falls but the form of the $s(Y_s)$ curve is retained, i.e., the parameters of the recombination centers are not altered and only their concentration is reduced. A different result was obtained by Rzhanov and Neizvestnyi [6] on heating of a germanium sample in liquid ethyl ether; this produced a large shift of the position of the $s(Y_s)$ maximum, indicating a change in the ratio of the hole and electron capture cross sections of the recombination centers. It is possible that this difference in behavior is related to the difference of the action of ether in the liquid and gaseous states. However, the reason for this difference is at present not clear.

The characteristic feature of the effect of ether vapor, distinguishing it from the other substances used in the present work, was the incomplete neutralization of the recombination centers generated by vacuum heating at 500°K. This result cannot be explained by the incompleteness of the adsorption process, because the vapor pressure of ether is very high (cf. Table 1) and the last measurements were carried out four days after the admission of ether vapor into the sample cell. Moreover, the experimental points fit well the straight line $s_{max} = a + bt^{-1/3}$, although they differ from the nitrobenzene results by the fact that the straight line does not pass through the origin of coordinates but intersects the ordinate axis at a point $s_{max} = 400$ cm/sec.

The results of the control experiment with benzene are given in Fig. 5. Thirty minutes after the admission of benzene vapor s_{max} had dropped by about 10% and only after two days did s_{max} reach half the value obtained after heating at 500°K. In the amyl alcohols and in chlorobenzene the recombination centers were almost completely neutralized after 1 hour. In the case of nitrobenzene and ethyl ether, two minutes after the admission of the vapor the surface recombination velocity was already reduced by a factor of about 2. Thus the control experiment with benzene shows that traces of water, present either on the walls of the working cell or in the vapors of the test liquids, cannot contribute significantly to the effects observed on the sample surface in the vapors of polar liquids.

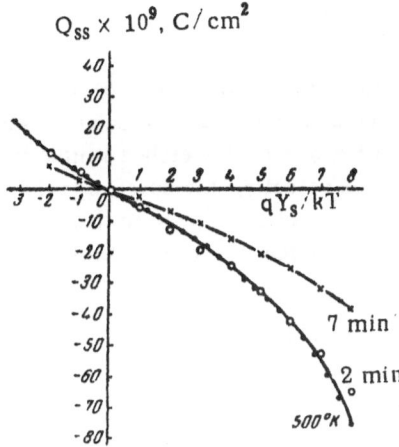

Fig. 4. Changes in the charge captured by the fast surface states on exposure to nitrobenzene vapor.

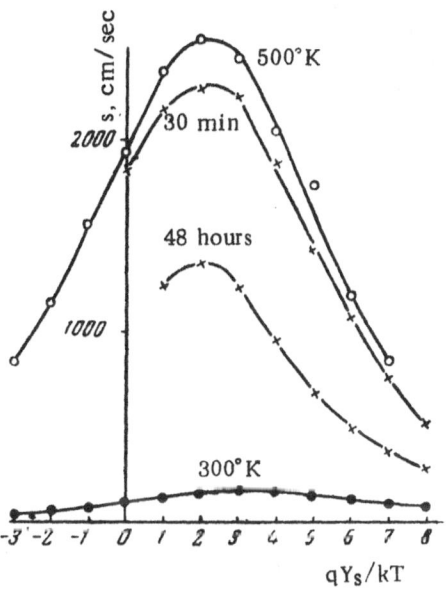

Fig. 5. Changes in the surface recombination velocity on exposure to an atmosphere of benzene vapor.

Discussion of Results

The results obtained for the absorption of molecules differing in their chemical structure allow us to conclude that physically adsorbed water plays the main role in the phenomenon of neutralization by water of the surface recombination centers. The process of neutralization is related to the establishment by a polar molecule of a local electric field which alters the parameters of a nearby recombination center. The electric field of the molecular dipole can, in principle, affect all the parameters of the recombination center: its energy position and the cross sections for the capture of holes and electrons.

Comparing the results given in Figs. 1 and 2 we conclude that the centers at which recombination and capture occur in the absence of polar molecules, lose first their recombination properties on approach of such molecules but continue to act as carrier-capture centers. Since the potential distribution of the charge captured by the fast surface states is not altered in the first 2 min after placing a sample in an atmosphere of nitrobenzene vapor, it follows that the energy positions of the centers do not change either. From this we may conclude that the electric field of a polar molecule which approaches a recombination center strongly alters the capture cross sections without greatly affecting the energy position of the center.

On prolonged exposure of a sample to nitrobenzene vapor the concentration of the capture centers begins after a time to fall. It is easy to see, however, that the $Q_{ss}(Y_s)$ curve then changes only in amplitude but retains its form, i.e., the relative distribution in energies of the capture centers remains the same. Moreover, the conductivity relaxation time of a germanium sample in a constant transverse electric field is strongly reduced by the adsorption of nitrobenzene molecules on its surface. In a sample heated in vacuum at 500°K the relaxation time is of the order of several hours; after several minutes in nitrobenzene vapor the relaxation time becomes several minutes, and finally, after 20 hours in benzene vapor it falls to several seconds. This means that the concentration of slow surface states is increased by the adsorption of nitrobenzene molecules on the germanium surface. A similar effect was observed earlier on heating of germanium samples in vacuum. Such heating not only increased the surface recombination velocity and the charge captured by the fast surface states but also increased the relaxation time of the slow states.

Considering these observations, it is logical to conclude that the electric field of a molecular dipole near a recombination center alters the carrier-capture cross sections of the latter so strongly that the center becomes a slow surface state.

The slowness of the neutralization of the recombination centers by nitrobenzene vapor shows that for this process to occur nitrobenzene molecules must diffuse through the oxide layer. On the other hand, the experimentally obtained power law for the time dependence of the maximum surface recombination velocity cannot

49

be deduced on the basis of diffusion alone. This obviously indicates that the process of neutralization is limited not only by diffusion but also by dipole locking near recombination centers.

Incomplete neutralization by ether vapor of the recombination centers produced by heating in vacuum may be related to some chemical reactions between ether and the germanium surface, which may alter the structure of the oxide layer. For example, it was found that immersion of a freshly etched sample in liquid ether for one hour increased several-fold the maximum recombination velocity. Moreover, a prolonged immersion in liquid ether frequently produced a strong macroinhomogeneity of the surface so that one side of the sample differed from another by about 0.1 V in its electrostatic surface potential. Consequently caution is necessary in interpreting experiments with ethyl ether.

The authors thank A. V. Rzhanov for his constant interest and valuable discussion of the results.

LITERATURE CITED

1. A. V. Rzhanov, Radiotekh. i Elektron. 1, 1086 (1956).
2. A. V. Rzhanov, N. M. Pavlov, and M. A. Selezneva, Zhur. Tekh. Fiz. 27, 2645 (1958).
3. A. V. Rzhanov, Yu. F. Novototskii-Vlasov, and I. G. Neizvestnyi, Fiz. Tverd. Tela 1, 1471 (1959).
4. A. V. Rzhanov, Yu. F. Novototskii-Vlasov, I. G. Neizvestnyi, S. V. Pokrovskaya, and T. I. Galkina, Fiz. Tverd. Tela 3, 822 (1961).
5. Yu. F. Novototskii-Vlasov and I. G. Neizvestnyi, Pribory i Tekh. Eksp. No. 4, 127 (1961).
6. A. V. Rzhanov and I. G. Neizvestnyi, Fiz. Tverd. Tela 3, 3317 (1961).

EFFECT OF ADSORPTION OF ETHER MOLECULES
ON GERMANIUM ON THE PARAMETERS OF
SURFACE RECOMBINATION CENTERS

I. G. Neizvestnyi

P. N. Lebedev Physics Institute, Academy of Sciences, USSR

In several earlier papers [1-5] it has been shown that the increase of the density of surface recombination centers produced by the action of ozone or the desorption of water can be completely annulled by placing the sample in contact with water vapor. This has led us to the conclusion that oxygen bound to the surface of germanium is an important component in surface recombination centers and that, on interaction with water molecules, these centers are neutralized, i.e., they cease to function as recombination centers.

We may assume that this effect is chemical. Then we can expect a water molecule to dissociate on the surface of germanium and either a proton or a hydroxyl radical to saturate the free valence of the recombination center, this valence being responsible for the surface recombination level. In such a case the neutralization effect should have two characteristic features. Firstly, it should be possible only with easily dissociated molecules which yield ions or radicals capable of saturating the free valence of the recombination center. Secondly, it is difficult to see how the parameters of the recombination center can change. The center can either function, with definite values of the energy level and effective cross sections for hole and electron capture, when the free valence is not saturated, or it cannot function as a recombination center when this valence is saturated.

The neutralization effect can be also electrostatic in nature with neutralization being caused by a considerable change in the recombination-center parameters, due to the action of a strong local field. The latter may be produced by an ion or a polar molecule located at a distance of the order of the lattice constant from the center. In this case we may find that the local field is insufficient for complete neutralization of the recombination center and that its action will produce only a quantitative change in the recombination-center parameters (energy position of the level, one or both capture cross sections).

The present work deals with a study of the neutralization effect of surface recombination centers on germanium.

Experimental Technique and Results

The purpose of the experiments was to investigate the recombination parameters of the surface centers on germanium when the adsorbed water molecules are replaced by other molecules. Measurements were carried out using the method of combining the field effect for strong sinusoidal signals with the steady-state photoconductivity, developed at our laboratory [7]. A germanium sample, located in a holder with mica spacers and transparent electrodes for the application of a transverse field, was immersed directly into the test liquid. For this purpose we used a special cell with a ground glass stopper into which molybdenum leads and a glass tube for circulating the thermostatting liquid were sealed. Constant temperature was maintained with an ultrathermostat and checked with a thermocouple placed near the sample. Optical glass was sealed into the bottom of the cell, through which the sample was illuminated with square light pulses during measurements of the photoconductivity.

In order to avoid leakage currents in the capacitor formed by the sample surface and the field electrodes, the liquid into which the sample is immersed must have a high electrical resistivity (higher than 10^{10} $\Omega \cdot$cm). This requirement limits very severely the number of suitable liquids, especially polar ones, the resistivity of which is as a rule low, and forced us to use only the nonpolar benzene and the weakly polar ethyl ether. Before use the test liquid was carefully purified in order to remove possible impurities, particularly water. The purification consisted of triple distillation, with selection of the middle fractions; during the first distillation a small amount of metallic sodium was introduced into the vessel with the liquid in order to dry it.

Before the measurements, the holder with the sample was kept for some time in the test liquid in order to establish an equilibrium state on the surface. Then the first curves of the dependence of the surface recombination velocity s and the charge captured on the surface Q_{ss} on the surface potential Y_{ss} were obtained at 300°K. Next, using a specially constructed automatic circuit, the sample was heated directly in the liquid by the current passing through it. After heating for 15 min the sample was cooled to the measuring temperature (300°K) and the same curves were again recorded. Such heating was repeated at several other temperatures up to 500°K.

The variation of the surface recombination velocity with the surface potential, $s = f(Y_s)$, after heating at various temperatures is shown in Fig. 1 for benzene and in Fig. 2 for ether. Simultaneously with changes in the surface recombination velocity there were changes in the charge captured by the fast surface centers. The appropriate data for benzene and ether before and after heating at 500°K are given in Figs. 3 and 4. Similar measurements were carried out on samples heated in ether before outgassing when the liquid layer still covered the sample and after outgassing, which desorbed the liquid molecules.

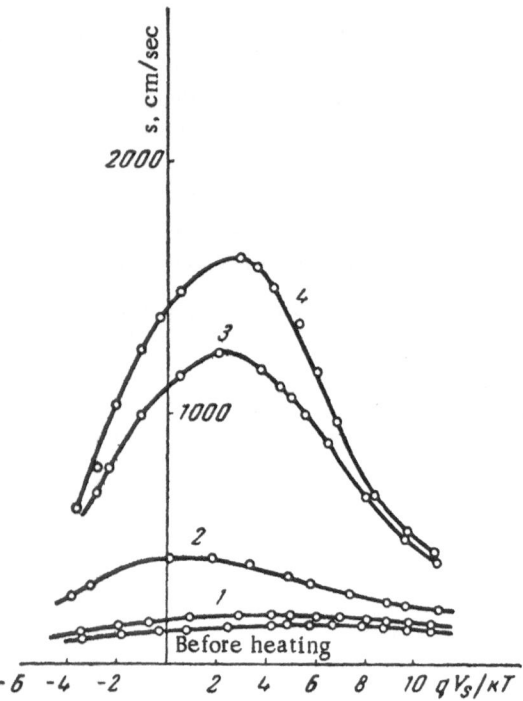

Fig. 1. Surface recombination velocity as a function of surface potential before and after heating at temperatures of 350 (1), 400 (2), 450 (3), and 500°K (4) in benzene. Measurements were carried out at 300°K.

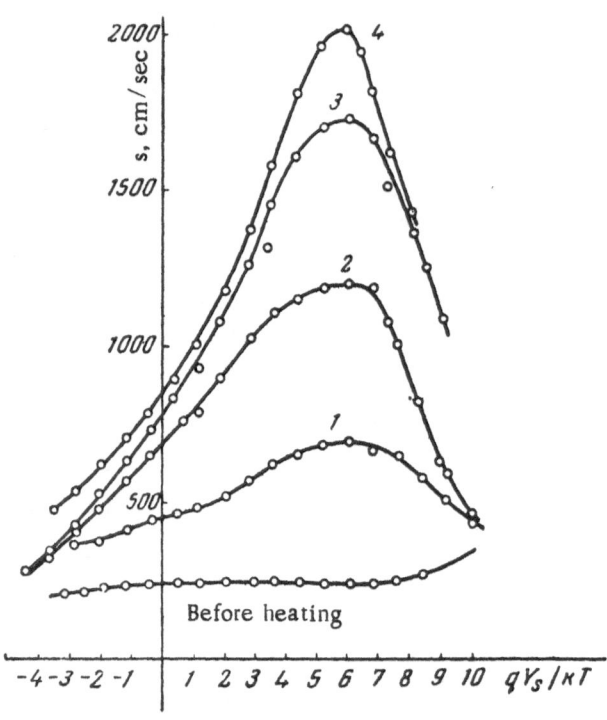

Fig. 2. Surface recombination velocity as a function of surface potential before and after heating at temperatures of 350 (1), 400 (2), 450 (3), and 500°K (4) in ether. Measurements were carried out at 300°K.

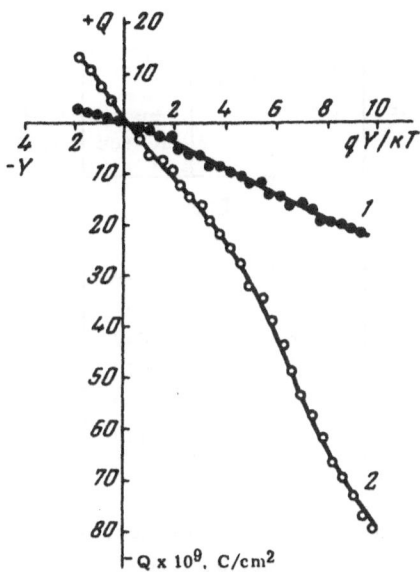

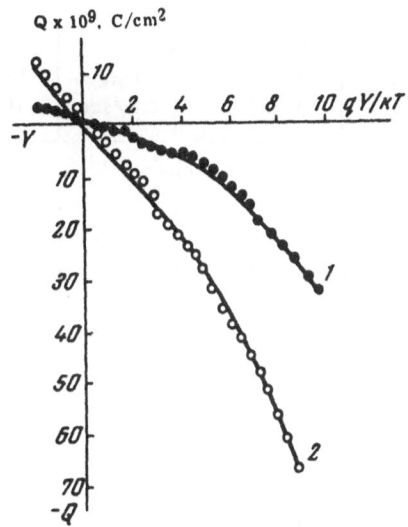

Fig. 3. Charge captured by fast surface centers as a function of surface potential: 1) before heating; 2) after heating at 500°K in benzene. Measurements were carried out at 300°K.

Fig. 4. Charge captured by fast surface centers as a function of surface potential: 1) before heating; 2) after heating at 500°K in ether. Measurements were carried out at 300°K.

The experimental results for five to six samples showed good qualitative agreement. Quantitative differences were found only in the maximum absolute values of the surface recombination velocity after heating. To illustrate these differences, we give a table which lists the maximum values of the surface recombination velocity, positions of these maxima and their half-widths for all the tested samples after heating at 500°K. These differences may be related to differences in the thickness and continuity of the oxide layers on the sample surfaces, which could have a pronounced effect since the samples were immersed in the test liquid immediately after etching. In all our other experiments the sample surface was usually stabilized in air or in vacuum for at least several days before measurements. The same influence of the thickness and continuity of oxide layers is obviously responsible for the strong surface heterogeneity of the samples heated in ether, the heterogeneity being particularly clear in field-effect experiments.

On carrying out tests on samples which were immersed in ether immediately after etching and placed in the holder while submerged in ether, we found no heterogeneity, although the samples were similar to others in the positions of the maxima on the curves which give the dependence of the surface recombination velocity on the surface potential. Since the heterogeneity did not appear, we may assume that heating in ether alters considerably the thickness of oxide layers on germanium.

Discussion of Results

If we compare the curves of the dependence $s = f(Y_s)$ obtained after heating in benzene with the corresponding curves after heating in vacuum, we see that they are completely identical. Both the positions of the maxima (near $+3kT/q$) and the half-widths of the curves, $(8-9)kT/q$, are the same. There are strong differences between the curves before heating. The curves representing the experiments carried out in vacuum have maxima, the curves representing tests in benzene increase monotonically. This is obviously related to different amounts of water adsorbed on the surface, since in the first stages of outgassing in vacuum the dependence $s = f(Y_s)$ is also monotonic. Similarly, the curves $Q_{ss} = f(Y_s)$ are practically identical for the two cases.

TABLE 1

Sample No.	Y_s (in kT/q) corresp. to s_{max}	s_{max}, cm/sec	Half-width (in kT/q)	Sample No.	Y_s (in kT/q) corresp. to s_{max}	s_{max}, cm/sec	Half-width (in kT/q)
Benzene				Ether			
44	2.5	1150	7.6	39	5.5	460	7.6
8	3.0	1600	8.8	35	5.3	1650	10.0
24	3.4	820	7.9	53	6.0	2000	7.2
54	3.2	1610	9.5	32	6.0	1070	10.0
53	3.0	1650	9.4	45	5.6	2100	8.8
				31	6.2	750	9.6

Thus we may conclude that the process of activation of the surface recombination centers occurs in benzene practically in the same way as in vacuum. This can easily be explained because the water desorbed on heating in benzene is replaced by nonpolar molecules of benzene and if these molecules are not chemisorbed on germanium, they should not affect the parameters of recombination centers.

On heating in ether, the process of activation of the surface centers is also analogous to heating in vacuum. If we take also into account the identity of the curves obtained before heating in ether and those obtained in benzene and in vacuum (during the initial stages of outgassing), we can say that ether by itself does not produce new surface recombination centers. On the other hand, as indicated by Figs. 1 and 2 and the table, the adsorption of ether displaces the maxima of the curves from a surface potential value close to $+3kT/q$, characteristic of benzene and vacuum, to a value near $+6kT/q$. Since the position of the maximum on the $s = f(Y_s)$ curve is determined by the magnitude of the ratio of the capture cross sections for holes and electrons, this shift indicates a considerable change in the parameters of the surface recombination centers or at least some of these centers. The observed shift of the maximum represents a change of the cross section ratio from ≈ 20 for vacuum and benzene to ~ 6000 for ether. In both cases we are dealing with acceptor-type centers since the cross section for hole capture is considerably greater than the cross section for electron capture, and the adsorption of ether increases this difference. Qualitatively, this result can be explained on the basis of Lax's idea [6], according to which large capture cross sections are related either to Coulomb or polarization effects of a center with a free charge carrier. The large hole capture cross section of a negatively charged filled acceptor center is due to the Coulomb interaction. It is not very likely that the adsorption of ether would alter considerably the nature of this interaction.

The large electron capture cross section of a neutral center is governed by the polarizability of the center in the field of the electron approaching it. In this case the adsorption of a polar ether molecule near a center may produce a strong polarization of the center, constant in direction, as a result of which its polarizability is strongly reduced. Consequently the electron capture cross section is reduced and the ratio of the cross sections for hole and electron capture increases, as found in our experiments.

We must consider the relationships between the maximum values of the surface recombination velocity in vacuum, benzene and ether. It is known that the capture cross section not only determines the position of the maximum on the $s = f(Y_s)$ curve but also the maximum value of the surface recombination velocity. In our case the hole capture cross section is unaltered but the electron capture cross section decreases by a factor of 300. According to the standard theory of recombination the magnitude of the maximum of the surface recombination velocity should decrease in direct proportion to the square root of the product of the cross sections, i.e., by a factor of 17. In our experiments the value of s_{max} for benzene and vacuum is the same, within the limits of experimental scatter, as the value for ether. This disagreement between theory and experiment can be very simply explained by the new relationship deduced for the surface recombination velocity by Rzhanov [8] in which both the numerator and the denominator depend in a different way on the square root of the capture cross-section product. Hence we may conclude that the maximum value of the surface recombination velocity should be practically unaffected by the changes in the capture cross sections.

In conclusion, we shall consider the data on the nature of the attachment of ether molecules to the germanium surface, obtained by transferring the holder with a sample from the liquid to the vacuum system. It was then found that as outgassing progressed the maximum on the $s = f(Y_s)$ curve gradually shifted from the position characteristic of ether to the position characteristic of vacuum. From this experiment we may finally conclude that the adsorption of ether molecules on the surface of germanium is a purely physical process and that their interaction with recombination centers is electrostatic in nature. In the case of the interaction of a center with a dipole molecule of water the electron capture cross section may be reduced so strongly that the center ceases to exist as a recombination center and becomes a "slow" state.

This work was carried out under the direction of Z. V. Rzhanov, to whom the present author is most grateful. The author is also indebted to G. A. Balandina who carried out the large amount of work necessary in purifying the liquids.

LITERATURE CITED

1. A. V. Rzhanov, Radiotekh. i Elektron. 1, 1086 (1956).
2. A. V. Rzhanov, Yu. F. Novototskii-Vlasov, and I. G. Neizvestnyi, Zhur. Tekh. Fiz. 27, 2440 (1957).
3. A. V. Rzhanov, N. M. Pavlov, and M. A. Selezneva, Zhur. Tekh. Fiz. 28, 2645 (1958).
4. A. V. Rzhanov, Yu. F. Novototskii-Vlasov, and I. G. Neizvestnyi, Fiz. Tverd. Tela 1, 1471 (1959).
5. A. V. Rzhanov, Yu. F. Novototskii-Vlasov, I. G. Neizvestnyi, S. V. Pokrovskaya, and T. I. Galkina, Fiz. Tverd. Tela 3, 822 (1961).
6. M. Lax, Phys. Rev. 119, 1502 (1960).
7. Yu. F. Novototskii-Vlasov and I. G. Neizvestnyi, Pribory i Tekh. Eksp. No. 4, 127 (1961).
8. A. V. Rzhanov, present collection, p. 70.

GAS EVOLUTION FROM THE SURFACE OF
SILICON

G. F. Romanova and I. I. Stepko

Institute for Semiconductors, Academy of Sciences, UkrSSR

Much work is currently being done on the surface of silicon. The interest in this problem is primarily due to the fact that the operation and reliability of silicon semiconducting devices depend to a considerable extent on their surface states.

Recent investigations have been mainly concerned with the determination of physical parameters (the topography and density of surface electron states) or with the study of various properties (including adsorption) of surfaces cleaned by various methods in vacuum. However, the physico-chemical structure of the real surface of silicon obtained immediately after etching is still not properly known.

The present paper describes an attempt to determine the amounts and compositions of gases adsorbed on the surface of silicon single crystals after etching.

Experimental Technique

Tests were carried out on p-type silicon samples of 50 $\Omega \cdot$cm resistivity. The sample dimensions were 25 x 3 x 0.6 mm. Before tests they were ground, etched in a mixture of 2HF and 3HNO$_3$ for 1 min, washed in distilled water, and dried with filter paper.

To expel the gases from the surface, the flash heating method was used. The duration of heating in different tests varied from 30 sec to 20 min. The sample temperature was measured with a Ni−Mo thermocouple (up to 600°C) and with an optical pyrometer. The uniformity of heating of the sample was checked by special tests because regions with different temperatures could have been a source of considerable error.

The gases evolved were analyzed in a glass mass-spectrometer system similar to that described by Pikus [1]. The sample was placed in a bulb connected by a wide passage to the mass-spectrometer tube. The apparatus is shown schematically in Fig. 1. During the flash heating a special ground-glass valve V was closed in order to reduce the rate of pumping (the pressure was halved in 5-6 min). The volume of the whole system was known: the pressure in the system during the flash, determined with an ionization gauge, was used as the measure of the amount of desorbed matter. After each measurement the system could be pumped to the initial pressure. The circuit for measuring the ion current allowed a mass spectrum to be recorded in 12-18 sec. Unfortunately, due to several secondary difficulties, the analysis of the spectrum was only qualitative.

It was found that the composition of the gases evolved from the sample surface was affected considerably by the heating of the glass system to 400°C (which was necessary to obtain high vacuum), and by the process of sealing-in the sample support into the test bulb. This can be understood by taking into account the fact that due to the evolution of large amounts of gases from the heated glass the residual gas composition in a high-vacuum system differs very considerably from the composition of a normal atmosphere [2], and the fact that during pumping, various reactions may take place on the sample surface, for example, the substitution of the substances initially adsorbed on the surface by the residual gases. Consequently, the pretest preparations were carried out as follows. Between a demountable bulb B (with a blank metal flange and a metal seal) and a mass spectrometer MS, a glass partition P was sealed-in before each test. The bulb without the sample was pumped for a considerable time by the pump P_1 and was heated in a furnace and in the flame of a burner. Then the bulb was opened, the prepared sample rapidly mounted in its holder (this operation did not take more than 15-20 min) and pumping continued for 15-18 hours but without heating the system. The prolonged continuous pumping made it possible to obtain a vacuum of 1×10^{-6} mm Hg. After reaching high vacuum a constriction C, well outgassed during the preliminary heating, was sealed (the pressure in the bulb did not increase then by more than half an order of magnitude) and the glass partition P was broken. The mass-spectrometer tube was by that time outgassed and cooled. The total pressure in the system was $(5-7) \times 10^{-7}$ mm Hg. This method of pumping deteriorated the initial vacuum and gave a somewhat more intense background in the mass spectrum. However, with this method of preparation for the tests, the time of contact of the sample with the gases evolved on sealing the bulb and on heating during pumping was reduced to a minimum.

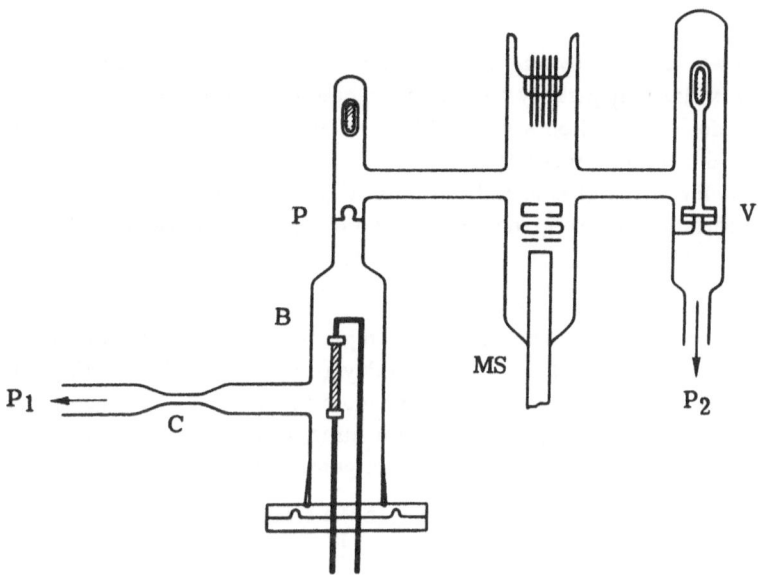

Fig. 1. Diagram of the vacuum system.

Results of Measurements

The variation of the pressure in the system at a given sample temperature is shown as a function of the duration of heating in Fig. 2. The readings were begun at the moment of switching on the heating current passed through the sample. The required temperature was established during the first few seconds (8-10 sec), as shown by curve A, and then the pressure rose rapidly by 3-4 orders of magnitude. This rise was so fast that it was impossible to record its kinetics by means of the usual vacuum gauge. If a series of successive brief heatings to the same temperature was carried out, the amount of gas evolved on each heating became less and less and, beginning with a certain heating, the pressure in the system assumed a steady value on switching on the current (after each flash the desorbed gas was pumped out). The total heating time during such a series of treatments did not exceed 3-5 min. The steady pressure indicated that the main part of the gas had been expelled from the surface.

After longer heating (up to one minute), the very rapid rise of pressure was followed by a drop of pressure, probably due to the pumping action of the ionization gauge and due to the adsorption on the cold parts of

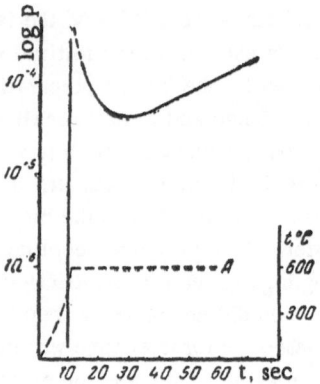

Fig. 2. Variation of the pressure as a function of the duration of heating.

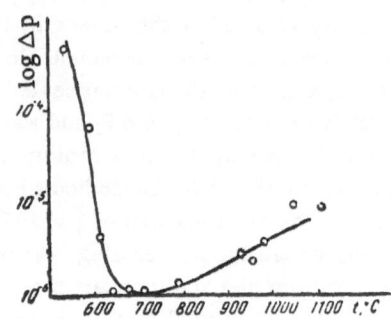

Fig. 3. Dependence of the amount of gas evolved from a sample on the sample temperature.

the system. The drop was followed by a new rise of pressure which was much slower and the causes of which are not clear. This slow rise may possibly be due to the desorption of molecules more strongly bound to the surface, or to the diffusion of gases from the interior of the sample, or to the decomposition of the surface oxide layer, though the second and third processes are not very likely at lower temperatures. The influence of stray factors, such as the heating of the glass or the evolution of gas from the molybdenum current leads, could not affect the results considerably (for the given construction of the bulb and the method of measurements) below sample temperatures of 1000-1100°C.

Each test sample was subjected to two cycles of measurements: the first, after etching, the second, after the vacuum heating, followed by the removal of the sample from the test bulb and the exposing of it to air for a shorter or longer time (from several hours to several days). During the first heating after etching, very large amounts of gases were evolved, corresponding to 20 or more monolayers. Such heating essentially altered the adsorption properties of the surface, while the second heating, after contact with air, desorbed only 2-3 monolayers. This result is similar to that obtained earlier for germanium [3].

Considering the dependence of the pressure rise in the system on the sample temperature, we see that the desorption from silicon occurs in two stages (the ordinate in Fig. 3 gives the total amount of gas evolved from the sample surface at a given temperature). A large amount of gas is evolved at temperatures up to 600-650°C, and then on further increase of temperature to 800°C the pressure is practically constant, but above 800°C there is another rise of pressure in the system.

The composition of the desorbed gas also varies with the temperature and the preliminary treatment of the sample. Figure 4 shows the mass spectrum of the gas evolved from the surface of a sample which was heated together with the whole vacuum system during pumping. Here, we see that the main component of the evolved gas has a mass of 28 and that the peak amplitude of this mass is two orders of magnitude greater than all the other peaks.

The sample which had not been exposed to gases evolved from the hot glass is characterized by a more varied spectrum in which the masses M_{18}, M_{28}, as well as M_2 and M_{44} are important (Fig. 5). The nature of the spectrum varies with temperature. Figure 6 shows the variation of the amounts of the separate components in the mass spectrum with the sample temperature: the M_{28} peak has the same two-step variation as the total pressure (Fig. 3); with the increase of the temperature to 700°C the M_{18} peak decreases considerably while the M_{44} peak increases together with the M_{28} peak.

During the second cycle the total amount of the evolved gas and the M_{18} peak decreased, while the M_{28} peak increased (Fig. 7).

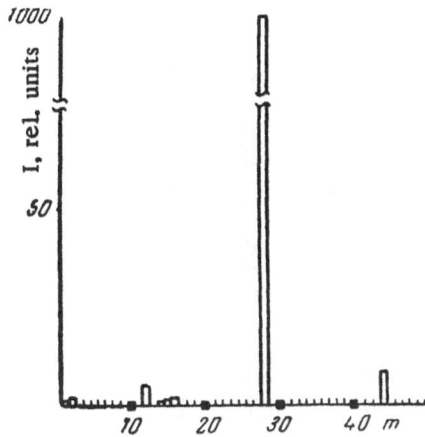

Fig. 4. Composition of the gases evolved from a sample which was heated together with the whole vacuum system (t = 600°C).

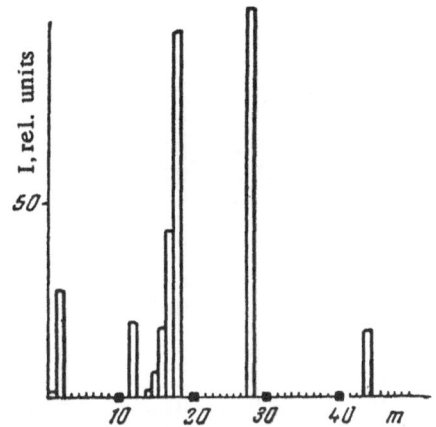

Fig. 5. Mass spectrum of the gases evolved from a sample which was cold during pumping (t = 510°C).

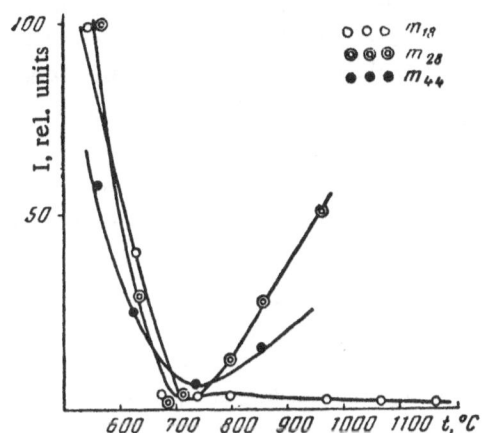

Fig. 6. Variations of the amounts of various components in the mass spectrum with the sample temperature (the sample was not heated during pumping).

The following experiment was also carried out: one of the samples was subjected to both test cycles and during the second cycle it was heated until the gas evolution stopped practically completely, and then it was etched again. The amounts and compositions of the gases evolved from this sample were the same as after the first etching (Fig. 8).

Discussion of Results

The results given above show that when the technique described here is used the gases are evolved mainly from the surface. The previous treatment of the sample affects strongly the composition and the amounts of the gases.

On analyzing the composition of the evolved gases, we conclude that they consist mainly of H_2O and CO. Comparison with the control spectra of N_2 and CO showed that nitrogen is not adsorbed at all on the etched surface of silicon or at best in very small quantities. Among other gases detected by mass spectrometry are H_2 and CO_2. However, in the presence of red-hot tungsten filaments in the system (the ionization gauge, the ion-source cathode) it is difficult to say whether the hydrogen is evolved from the sample surface or whether it is the product of the decomposition of water vapor on tungsten.

If the sample is heated together with the whole system in the process of obtaining high vacuum, the greatest difference is observed in the composition of the gases desorbed at lower temperatures: H_2O disappears completely and more than 90% of the gas is composed of CO. This is in full agreement with the results of Wolsky and Zdanuk [2] on the residual gases in high-vacuum systems; they found that the main component is CO. The substances initially adsorbed on the Si surface are probably replaced by the gases evolved from the glass during heating. The possibility of such a substitution was pointed out by Maxwell and Green [4].

If the sample is not in contact with hot glass, its surface retains a very thick and complex film which is formed after etching. The desorption of this film explains the more complex mass spectrum obtained for etched samples. The complexity of the structure of the film on the Si surface is indicated by the two-stage

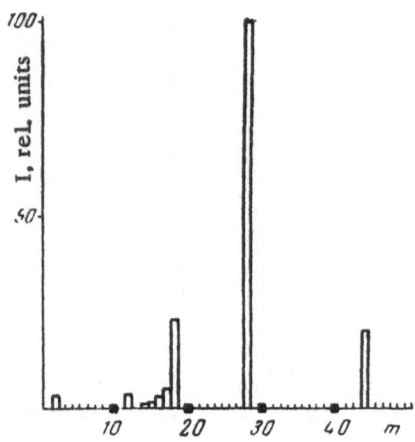

Fig. 7. Mass spectrum of the gases evolved during the second cycle of measurements (after contact with air; t = 500°C).

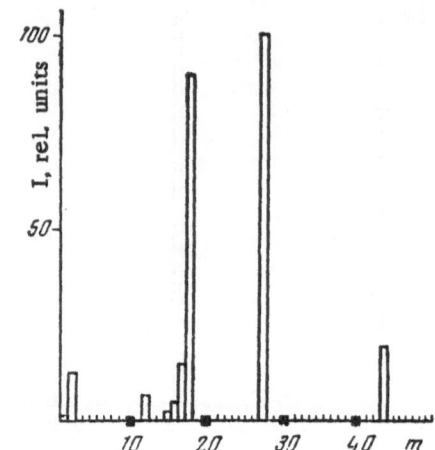

Fig. 8. Mass spectrum of the gases evolved from an etched sample (t = 600°C).

evolution of gas: at temperatures below 600-650°C the relatively weakly bound adsorbed substances are removed from the surface, while the gas evolution increasing above 800°C is probably related to the decomposition of the more strongly bound oxide layer or to the removal of chemisorbed substances. Obviously, oxygen bound to the surface is evolved as CO and CO_2.

After heating, even if it is not long, to a high temperature (above 800°C) the adsorption properties of the surface are altered, which is indicated by the second cycle of measurements. The change in the amount of desorbed matter is so great that it cannot be accounted for, for example, only by the change in the roughness of the sample surface.

Experiments on etched samples confirm that it is the etching of the Si surface that produces a thick layer, including a large amount of water, and representing a complex structure of oxides, hydroxides and the gases adsorbed on them. The existence of such a layer on germanium has been reported by Ellis [5]. Heating destroys this layer, and the subsequent formation of an oxide layer and the adsorption on silicon on contact with air are quite different in nature from the processes occurring after etching.

Few papers have been published on gas evolution from silicon; these kinds of studies were mostly carried out for the purpose of finding the role of gases dissolved in the interior [6]. The work of Kozlovskaya [7] on the gas evolution from a silicon surface has not, in our opinion, allowed for several important factors: the contact of the sample with the gases evolved from the hot quartz; the quartz bulb outgassed at a higher temperature was cooled somewhat during the test and could itself have absorbed a considerable amount of gas; the water expelled from the sample could have been decomposed on the hot quartz so that the amount of hydrogen would have been grossly overestimated.

In conclusion, we stress once again that the results obtained allow us to assert that the real etched surface has a very complex structure and that further studies of the structure and the chemical composition of the surface film should help in ascertaining the nature of the surface electron states.

We are very grateful to V. I. Lyashenko for constant interest in this work and discussion of the results.

LITERATURE CITED

1. G. E. Pikus, Izvest. Akad. Nauk SSSR, Ser. Fiz. 20, 1080 (1956).
2. S. P. Wolsky and E. J. Zdanuk, Vacuum 10, 13 (1960).
3. I. I. Stepko, G. F. Romanova, and N. G. Chmel', Ukrain. Fiz. Zhur. 5, 5 (1960).
4. R. H. Maxwell and M. Green, J. Phys. Chem. Solids 14, 94 (1960).
5. S. G. Ellis, J. Appl. Phys. 28, 1262 (1957).
6. H. A. Papazian and S. P. Wolsky, J. Appl. Phys. 27, 1561 (1956); J. H. Crawford, J. Appl. Phys. 27, 839 (1956).
7. M. V. Kozlovskaya, Fiz. Tverd. Tela 1, 1027 (1959).

EFFECT OF ELECTROCHEMICAL TREATMENT
ON THE SURFACE RECOMBINATION VELOCITY
OF GERMANIUM IN VARIOUS GASEOUS
ENVIRONMENTS

G. V. Smirnov, Yu. M. Polukarov, and
V. A. Arslambekov

Physical Chemistry Institute, Academy of Sciences, USSR

Several workers [1-7] have shown that adsorbed gases and water vapor, films representing separate phases, and other contaminations present on a germanium surface strongly affect the absolute values of the electrical parameters of germanium, as well as the reproducibility and the stability of these parameters. In this connection, it is of interest to combine the results of studies of the physico-chemical properties and of the electrical parameters of a germanium surface.

The present paper presents the results of studies of the surface recombination velocity s in various gaseous media, carried out on samples subjected to various chemical and electrochemical treatments. Simultaneously, using various methods, studies were made of the mechanism of the influence of the physico-chemical state of a germanium surface on its recombination velocity.

Method

The work was carried out mainly on single-crystal n-type germanium samples of resistivity $\rho = 10$ $\Omega \cdot$cm and diffusion length $L = 1.2$ mm, cut parallel to the (111) plane and having the dimensions $20 \times 5 \times 0.15$ mm. The results obtained for these samples did not differ essentially from the results for n-type germanium samples of 40 $\Omega \cdot$cm resistivity and 2.5 mm diffusion length.

After the samples were ground and ohmic contacts soldered to them, they were etched in hydrogen peroxide ("very pure" grade) and washed in hot (80°C) distilled water. Then the surface recombination velocity was determined from the decay of the photoconductivity.

Apart from this normal washing, the samples were also washed in distilled water with simultaneous anodic polarization at various potentials and temperatures (the potential and temperature were fixed for a given sample). Platinum plates placed on both sides of the germanium served as the cathodes. The potential of the germanium anode was measured with respect to a saturated calomel electrode using a Luggin capillary and an electrometer amplifier type EMU-3 (input impedance not less than 1.5×10^9 Ω).

In many cases the germanium surface was deliberately contaminated with copper by immersion in a solution of copper nitrate.

Measurements of the surface recombination velocity were carried out under the following conditions: atmospheric air, vacuum of 10^{-6} - 10^{-7} mm Hg, heating at 120°C and return to the previous vacuum, dry oxygen, atmospheric air. Special tests showed that the contact of a sample with grease vapors strongly alters the surface recombination velocity. Consequently, all measurements in vacuum and dry oxygen were carried out in a glass tube which was connected to a vacuum system through a trap cooled with liquid nitrogen. The durations of exposure of the samples to the above environments listed above varied from several hours to several hundreds of hours. In the majority of cases, we investigated the variation of the surface recombination velocity with time.

Results and Discussion

Normally, there are adsorbed layers of gases and water vapor on a germanium surface. Consequently, it is very important to know the amounts and compositions of these layers and the mechanism of their influence on the electrical properties of germanium.

We investigated the kinetics of adsorption and desorption of water vapor and gases on germanium surface, using a very sensitive precision vacuum quartz damped balance (sensitivity 1×10^{-7}g, load up to 7 g, time to come to complete rest 15 sec) [8], shown in Fig. 1. This investigation showed that during the exposure of a germanium sample to a humid atmosphere, a large amount of moisture (up to 20 μg/cm^2, representing a water layer of up to 2000 A thickness) is adsorbed on the surface. The amount of adsorbed moisture depends markedly on the previous history of the sample — in particular on the thickness and composition of the oxide layer, on the pressure and duration of action of the water vapor, and on several other factors.

To remove only half the adsorbed moisture from a germanium surface it is necessary to keep the sample in 10^{-6} - 10^{-7} mm Hg vacuum for several tens of hours. This time is shortened to several hours by vacuum heating at 120-200°C.

For a germanium sample to reach constant weight on heating in vacuum at temperatures up to 200°C several tens of hours are needed.

From the data on the kinetics of adsorption and desorption of water vapor in the presence of an oxide layer on a germanium surface, we may conclude that water vapor reacts chemically with the surface oxide layer, apparently forming a hydroxide layer.

The results obtained using the quartz balance method agree with those of an investigation of the surface recombination velocity in vacuum. If a sample of germanium is etched and washed in the normal way, kept for about 20-30 hours in air (until an approximately constant value of s, of the order of 40-100 cm/sec, is reached) and then placed in vacuum, the value of s increases and the longer the sample is kept in vacuum the greater this increase (Fig. 2). Several tens of hours are needed in high vacuum for the surface recombination velocity to become constant having reached values of 600-800 cm/sec. Such a slow variation of s is related to the relatively slow process of water desorption from the germanium surface.

Heating the sample in vacuum for several hours at 120°C produces a further increase of s to 1200-1400 cm/sec due to the further removal of water from the germanium surface.

On the subsequent admission of dry oxygen, a practically constant value of the surface recombination velocity is established in several minutes. At an oxygen pressure of 15 mm Hg, this value is 2300-2500 cm/sec.

Subsequent prolonged storage of the sample in vacuum re-establishes the s values observed directly before heating in vacuum. This fact, together with a special study (by the balance method) of the oxygen adsorption and desorption processes, indicate that the adsorption of oxygen on an oxidized germanium surface is reversible. If atmospheric air is then admitted into the system, after many tens of hours s approaches the values which were observed for the same samples after washing and exposure to air.

After preliminary treatments of the germanium surface (washing under anodic polarization conditions, contamination of the surface with copper), the nature of variation of s with time in the listed gaseous environments was similar to the variation described above.

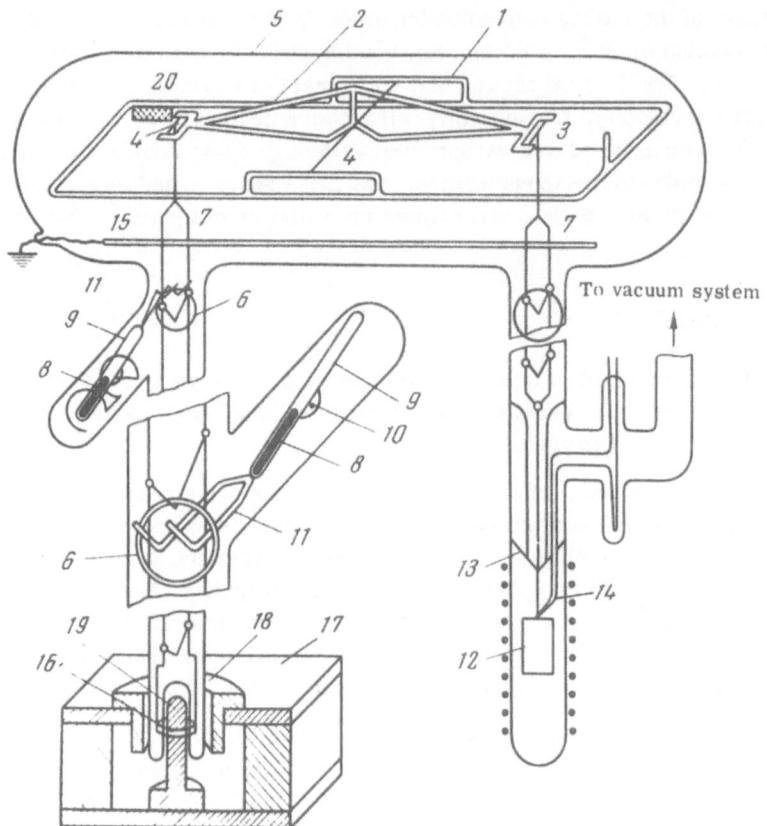

Fig. 1. General appearance of the balance. 1) Frame; 2) arm; 3) pointers; 4) suspension filaments; 5) bulb enclosing balance; 6) weights; 7) stirrups; 8) iron core; 9) glass tube; 10) axle; 11) lifting hooks; 12) sample; 13) screen; 14) thermocouple; 15) arresting device; 16) damping ring; 17), 18), and 19) magnetic system; 20) indicator.

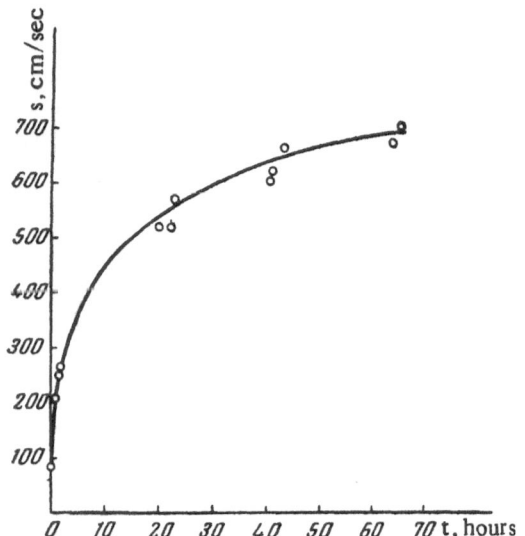

Fig. 2. Dependence of the surface recombination velocity of n-type germanium on the duration of storage in vacuum. After etching and washing the sample was kept in air for 26 hours and then placed in vacuum.

If etched germanium samples are washed in distilled water under conditions of anodic polarization, the values of the surface recombination velocity in all the investigated gaseous environments are considerably lower than in the case of normal washing in water of the same purity. Thus, after washing in hot distilled water at potentials of +(0.3-0.6) V applied to germanium, the following values of s were obtained: 20-40 cm/sec in air; 40-80 cm/sec after long storage in vacuum; in dry oxygen s was also several times smaller than s in dry oxygen after the normal washing in distilled water.

It is worth noting that the reproducibility of the s values is considerably better after washing in distilled water under anodic polarization conditions than after normal washing.

The lower and more reproducible values of the surface recombination velocity of germanium samples subjected to washing under anodic polarization conditions may be due to the fact that this method of washing removes from the germanium surface the metallic and other contaminations which have settled on this surface during the preliminary treatments (grinding, etching, etc.). Moreover, it is also possible that the anodic-polarization washing orders the structure of the surface layers of germanium and its oxide, in addition to altering the composition, structure and thickness of the oxide.

To find whether it is possible to clean a germanium surface of metallic contaminations, we carried out tests which included deliberate contamination of the surface with copper followed by anodic washing at various potentials. The samples were immersed for several minutes in a hot solution of copper nitrate containing $1 \times 10^{-4}\%$ copper and then s was measured in the aforementioned gaseous environments. Typical values of s were as follows: 600 cm/sec in air; 1100 cm/sec in vacuum; 2600 cm/sec in vacuum after heating at 120°C. Then the samples were washed under anodic polarization conditions using a fixed potential. Some results of the measurements of s carried out after anodic washing in hot distilled water, are given in Fig. 3, where s is shown as a function of the potential applied to the germanium sample during washing. The same figure gives the results of measurements of s in vacuum after long storage in that medium (curve 1) and after heating in vacuum at 120°C (curve 2).

It is significant that anodic washing in distilled water of samples contaminated with copper produced approximately the same values of s in all the investigated environments as did the same type of washing immediately after etching. This, and the low and reproducible values of s, may be regarded as a proof that the anodic washing removes copper and other harmful contaminations from the germanium surface. This cannot be said of the normal washing.

Experiments carried out on the washing of etched germanium samples in tap water under anodic polarization conditions (as well as experiments on the normal washing in tap water) showed that the values of s obtained after such washing are several times greater than the values after washing in distilled water with simultaneous polarization.

In order to study the structure of the surface oxide layers on germanium, we subjected to electron diffraction the samples which were etched and washed in distilled water in the normal way or under anodic polarization conditions. Unfortunately, these tests did not give any information on the structure of the oxides on the germanium surface because, apart from the reflections representing the crystal lattice of germanium, no other reflections were found. The question of the existence of oxide layers on germanium etched in hydrogen peroxide and washed by either of the two methods described previously is still unresolved. All we can say is that on the samples tested, thick oxide layers were not formed.

To investigate in greater detail the influence of various types of preliminary chemical or electrochemical treatment on the surface recombination velocity, the latter was measured not only at room temperature but also at temperatures from 20 to 70°C. To avoid the complications introduced by the desorption of water vapor during the study of the temperature dependence of s, the germanium samples were subjected to preliminary heating in vacuum at 120°C for several hours and then s was measured in vacuum and in dry oxygen in the 20-70°C range. The dependence of the logarithm of the surface recombination velocity on the reciprocal of the temperature (Fig. 4) indicates the strong effect of the contaminations, present on the germanium surface, on the value of s. The results of Fig. 4 indicate the possibility of cleaning

Fig. 3. Dependence of the surface recombination velocity of germanium on the anodic-polarization potential during washing in distilled water at 80°C. Before this anodic treatment, the samples were contaminated with copper. 1) Vacuum before heating; 2) vacuum after heating. The arrow indicates the steady-state potential of germanium in the absence of an external field.

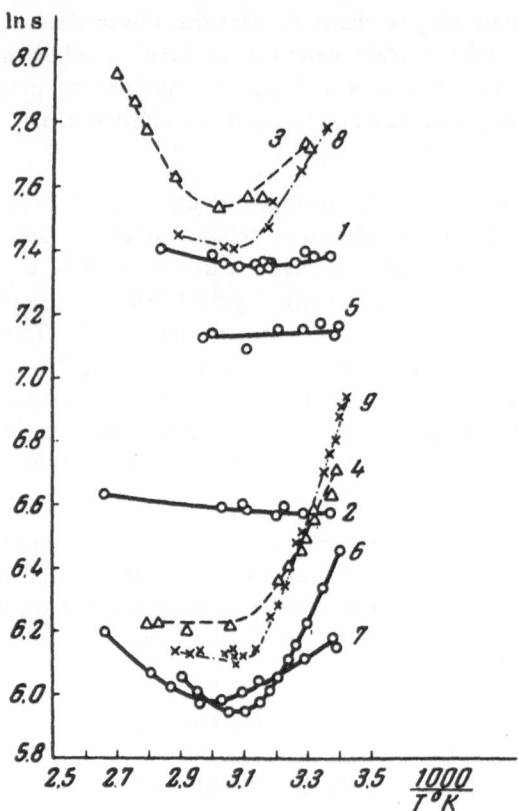

Fig. 4. Dependence of the logarithm of the surface recombination velocity on the reciprocal of the test temperature. 1)-7) Measurements in vacuum after heating in vacuum; 8) and 9) measurements in dry oxygen after heating in vacuum (15 and 26 mm Hg, respectively).

the germanium surface from contaminations by washing and simultaneous anodic polarization. Curves 8 and 9 in Fig. 4 represent the result of measurements in dry oxygen, the other curves refer to vacuum measurements. Curves 1 and 8 represent a sample which was washed after etching in the normal way in hot distilled water; curves 2 and 9 represent the same sample subjected, after the normal washing and measurement of s, to the anodic washing at a potential of 0.24 V in water of the same purity and temperature as used to obtain curves 1 and 8. Curves 2 and 9 show the undoubted advantage of washing under anodic-polarization conditions over the normal washing in water of the same purity and temperature.

Curve 3 represents a sample which was etched and then washed for several minutes in a hot solution of copper nitrate, containing 1×10^{-4} % copper; curve 4 shows the results for an identical sample which, after etching, was washed in the same solution of copper nitrate but an anodic potential of 0.24 V was applied to germanium during this washing. These two curves indicate that copper was deposited on the germanium surface in the first case but in the second there was either no or very little deposition.

Curves 5, 6, and 7 were obtained for samples which were contaminated with copper and then subjected to anodic polarization in hot distilled water at potentials of 0, 0.44, and 0.54 V, respectively. After such treatment much lower values of s were obtained compared with the values obtained for the accidentally (in the process of etching and washing in water) and the deliberately contaminated germanium samples.

Attempts were made to explain the reason for the differences in the temperature dependence of s. However, so far no definite correlation was found between the nature of the temperature dependence of s on the one hand, and the chemical or electrochemical treatment, nature of the gaseous medium, and absolute value of s on the other.

In many cases, a very weak dependence of s on temperature was observed. This was found both for the normal washing (curve 1 in Fig. 4) as well as for the washing with simultaneous anodic polarization at not-too-high germanium potentials (curves 2 and 7).

It follows from the comparison of the temperature dependence in vacuum and in oxygen, that the admission of the latter alters in many cases both the absolute value of s and its temperature dependence.

The authors regard it as their pleasant duty to thank Professor Doctor of Chemical Sciences K. M. Gorbunova for her interest in this work and her valuable advice.

LITERATURE CITED

1. W. Brattain and J. Bardeen, Bell System Tech. J. 32, 1 (1953).
2. A. Stevenson and R. Keys, Physica 20, 1041 (1954).
3. S. Wang and G. Wallis, Phys. Rev. 107 947 (1957).
4. S. Ellis, J. Appl. Phys. 28, 1262 (1957).
5. M. Lasser, C. Wysocki, and B. Bernstein, Phys. Rev. 105, 491 (1957).
6. V. I. Fistul' and D. G. Andrianov, Doklady Akad. Nauk SSSR 130, 374 (1960).
7. S. Morrison, J. Phys. Chem. Solids 14, 214 (1960).
8. V. A. Arslambekov, Highly Sensitive Vacuum Quartz Damped Microbalance, TsITEIN, Advanced Scientific Technical and Industrial Experience, Subject 29, No. II-60-72/ 9 (Moscow, 1960).

II

ELECTRICAL PROPERTIES OF SEMICONDUCTOR SURFACES

INVESTIGATION OF SOME ELECTRON PROCESSES ON A REAL SURFACE OF GERMANIUM

A. V. Rzhanov

P. N. Lebedev Physics Institute, Academy of Sciences, USSR

1. Introduction

The development of combined experimental methods, involving the simultaneous investigation of changes in the surface conductivity, surface recombination and carrier capture under the action of electric fields perpendicular to a semiconductor surface, brought about a great advance in the understanding of these electron processes on the surfaces of semiconductors.

Several variants of such combined methods have been developed in the Laboratory for Semiconductors of the P. N. Lebedev Physics Institute of the USSR Academy of Sciences, and over the years systematic studies of the electron processes on the surface of germanium have been carried out. The experimental technique and the treatment of the results were described in a series of original papers [1-5] and we shall not deal with them here.

The present communication is devoted to a general analysis of the currently known experimental facts and relationships governing surface recombination and capture on the so-called real surface of germanium and to a comparison of these data with theoretical representations. It is assumed that the surface scattering of carriers is completely isotropic under experimental conditions so that the carrier mobility near the surface can be calculated by means of the theoretical relationships of Schrieffer [6] or the more exact work of Greene et al. [7]. On this assumption, the data on the surface conductivity can be used to determine directly the surface potential values corresponding to various instantaneous values of the applied transverse electric field. Consequently, most of the experimental data being analysed is in the form of dependences of the surface recombination velocity and of the charge captured at the surface on the surface potential.

The first experimental problem is to ensure sufficient reproducibility of the state of the semiconductor surface, which can be judged by the reproducibility of the nature of the experimental dependences referred to above.

The (labor-consuming) standardization of all the stages of surface treatment and preparation for measurement, as well as the use of 10^{-6} mm Hg vacuum as a standard atmosphere during measurements, solved this problem. To illustrate the degree of reproducibility of the experimental results, Fig. 1 shows the dependence of the surface recombination velocity on the surface potential obtained for two different samples (300° K). The differences between the relative values of the surface recombination velocity of the two samples are so small that a common curve can be drawn for both of them. The differences between the absolute values of the surface recombination velocity are greater at the maximum but they are still within the limits of experimental error. Similarly, the differences between other properties of the test samples are well within the limits of

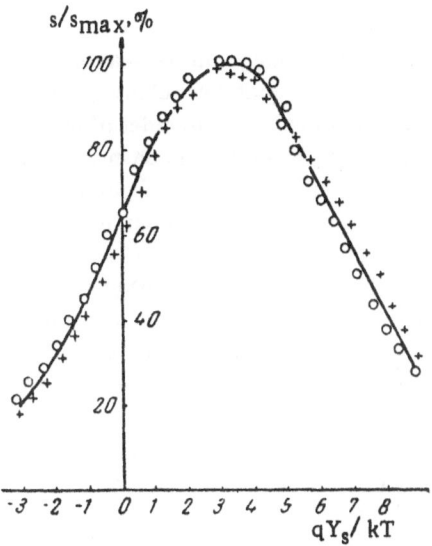

Fig. 1. Dependence of the surface recombination velocity (in relative units) on the electrostatic surface potential: $\bigcirc$ - s_{max} = 250 cm/sec; + - s_{max} = 290 cm/sec.

experimental error. Among these properties are: the charge captured by the fast surface states as a function of the surface potential, and the total surface charge in the absence of an external transverse field.

The achieving of the reproducibility of the experimental data establishes a reliable basis for their analysis. The nature of the experimental curves clearly indicates participation of several types of center, with different parameters, in the processes of recombination and capture on the surface of germanium. It would have been possible to determine these parameters if the experimental recombination and capture curves could be separated into basic curves corresponding to monoenergetic discrete levels or to systems of levels distributed according to some law in the forbidden band at the surface. Analysis shows, however, that such a separation cannot be made unambiguously because the experimental curves have no characteristic features which would justify the selection of a particular method of separation.

Therefore, to identify the systems of surface recombination and capture centers it is necessary to use additional experimental data. One would expect to obtain such data by investigating the influence of various external agencies on the processes of surface recombination and capture. The following were selected as the agencies: variation of the injection level; variation of temperature; agencies altering the concentration of surface centers; agencies altering the properties of surface centers.

The results of such investigations allowed us to draw several conclusions about the nature of the system of surface recombination and capture centers; they were supplemented by the result of studies of recombination and capture for different crystallographic orientations of the sample surfaces, and of the spectral distribution of the photoconductivity generated by the excitation of surface levels. These results made it possible, first of all, to prove the invalidity of the popular assumption of the dominance of one type of recombination center governing the value of the surface recombination velocity at its maximum. This assumption was clearly contradicted by the experimental data on the dependence of the maximum values of the surface recombination velocity on the injection level [8] and on temperature [9], by the dependence of the position of the recombination-curve maximum on temperature [9] and, finally, by the relationships between the changes in the maximum values of the surface recombination velocity and the slopes of the capture curves in experiments involving variation of the recombination center concentration [10]. The latter experiment deserves a more detailed discussion.

2. Experiments on Variation of the Concentration of Recombination and Capture Centers

Quite a long time ago it was established that the action of ozonized oxygen and heating of samples in vacuum to moderate temperatures, not exceeding 700°K, both produce considerable changes in the concentrations of surface recombination and capture centers [1-3]. These experiments enable us to draw some conclusions on the nature and structure of surface recombination and capture centers and we shall deal with this later. At the same time the results of these experiments, particularly on the effect of heating in vacuum, are of considerable interest from the point of view of interpretation of the experimental recombination and capture curves. Typical results for one sample are shown in Figs. 2 and 3. All the curves were obtained at the same test temperature of 300°K and the figures by the curves represent the temperatures of heating before tests.

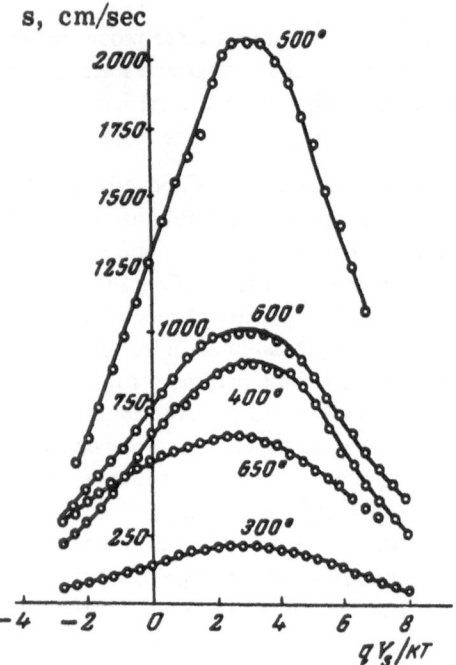

Fig. 2. Dependence of the surface recombination velocity on the electrostatic surface potential obtained at $300°K$ in 1×10^{-6} mm Hg vacuum. The figures by the curves give the temperatures of heating before measurement (in $°K$).

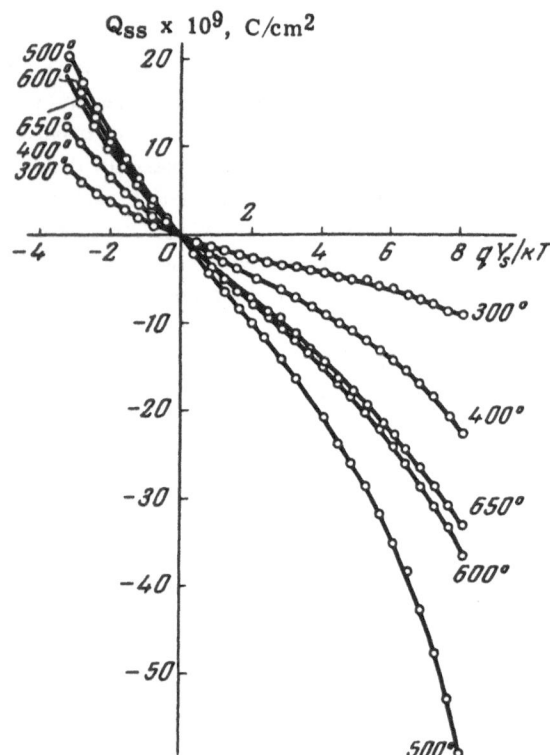

Fig. 3. Dependence of the charge captured by the fast surface states on the surface potential. The test conditions and notation are the same as in Fig. 2.

Figure 2 shows that such heating produces very considerable changes in the maximum values of the surface recombination velocity, these changes ranging over a whole order of magnitude. However, in spite of such large changes in the absolute values, the shape of the recombination curves remains the same. This can be seen by comparing the reduced $s/s_{max} = f(Y_s)$ curves, the differences between which lie within experimental error. This means that either the concentrations of different types of recombination center vary practically identically with heating, or the nature of the sum of the contributions of various centers is such that the resultant curve is not very sensitive to changes in the ratios of the concentrations of various centers.

Changes in the slopes of the capture curves due to heating (Fig. 3), which represent changes in the concentrations of surface centers, do indeed vary more or less identically throughout that part of the forbidden band at the surface which was investigated in these tests. A qualitative correlation is observed between the sign of the change of the general slope of the capture curves and the sign of the change of the maximum values of the surface recombination velocity, both changes being produced by heating to various temperatures. If the surface recombination velocity in the region of its maximum is governed by recombination centers of one type, one would expect direct proportionality between the values of the slopes of the capture curves at some value of the surface potential and the maximum velocities of the surface recombination.

Comparison of the data in Figs. 2 and 3 shows that in fact not only is there no such direct proportionality but even a quantitative correlation is lacking. Thus, for example, the maximum surface recombination velocity is considerably greater after heating at $400°K$ than after heating at $650°K$ although the ratio of the slopes of the corresponding capture curves has always the opposite trend. The slopes of the capture curves after heating at 600 and $650°K$ are very close to one another while the maximum values of the surface recombination velocity after such heatings differ by a factor of nearly two. Conversely, the recombination curves after heating at 600 and $400°K$ differ very slightly by a quantity which is practically independent of the surface potential, while the slope of the corresponding capture curves are very different.

These discrepancies are the strongest proof of the invalidity of the assumption that the surface recombination velocity near its maximum is governed by centers of one type, the concentration of which can be determined from the slope of the capture curve. An analysis of these discrepancies leads to the conclusion that among the components of the surface recombination velocity there should be one weakly dependent on the surface potential but varying strongly under the action of external agencies, such as heating in vacuum.

The actual existence of such a surface recombination velocity component is confirmed by a whole series of direct experiments. Thus, for example, this component obviously dominates surface recombination on freshly etched samples in an insufficiently dried atmosphere or even in vacuum during the first stages of pumping. Figure 4 gives the data on the change in the nature of the recombination curve with decrease of the humidity and the pressure in a closed system containing the sample, which illustrates well the change in the relationships between the various components of this surface recombination velocity.

The component of the surface recombination velocity which is weakly dependent on the surface potential is undoubtedly related to levels continuously distributed in their energy positions in the forbidden band. However, the usual theory of recombination always gives for these levels a very narrow recombination curve, the width of which is governed by the distribution law of the levels. At the same time, it is very likely that the densities of these levels, which, according to the capture data, increase rapidly away from the middle of the forbidden band, are sufficient for coalescence into surface bands. If these surface bands are in sufficiently good electrical contact with the volume conduction and valence bands, their existence is equivalent to a reduction of the forbidden band width at the surface. The probability of the direct radiative band—band recombination increases then at the surface and the velocity of this recombination depends weakly on the surface potential but varies very rapidly with the change in the surface level density.

An experimental verification of the proposed mechanism can be obtained by investigating recombination radiation from the surface of germanium.

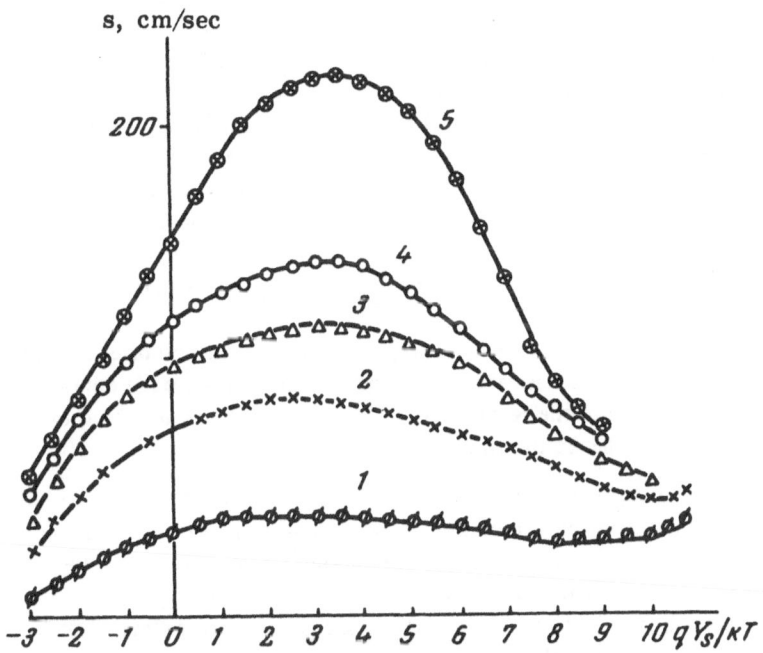

Fig. 4. Successive changes in the nature of the dependence of the surface recombination velocity on the electrostatic surface potential when the humidity and pressure of the ambient atmosphere are reduced. The test temperature was $300°K$. 1) Atmospheric pressure; 2) atmospheric pressure with liquid nitrogen in a vacuum trap; 3) 2×10^{-2} mm Hg; 4) 5×10^{-5} mm Hg; 5) 2×10^{-6} mm Hg.

3. Nature of Surface Recombination and Capture Centers

Experiments on the variation of the density of surface recombination and capture centers led to some conclusions about their nature. The basic facts were the practically identical effects of ozonized oxygen and vacuum heating of a germanium surface on the processes of surface recombination and capture.

It follows that defects in germanium—oxygen bonds are the chief components of recombination and capture centers, and that the adsorption of water molecules near the centers neutralizes their recombination action. The validity of the latter conclusion was confirmed by direct experiments in which changes in the surface state, produced by vacuum heating or by ozonized oxygen, were completely neutralized by placing the sample in contact with a humid atmosphere.

Of great interest was the nature of the neutralization of recombination centers by the adsorption of water molecules on germanium. To solve this problem, several experiments were carried out, described in the papers of Yu. F. Novototskii-Vlasov and P. G. Neizvestnyi,[*] in which adsorbed water molecules on a germanium surface were replaced by various molecules of other compounds. Without analyzing in detail the results of these experiments, we shall only say that they proved the electrostatic nature of the interaction between a recombination center and an adsorbed molecule, which produces a strong reduction of the effective capture cross section of carriers of one type. In this process, the center is transformed from a fast recombination state into a slow state which does not take part in the processes of recombination and capture in these experiments.

The interpretation of the neutralization effect is closely related to another problem: the change in the total surface charge due to heating in vacuum [11]. The experimental technique used to study this problem allowed us to determine directly the value of the total surface charge and to follow its changes under the action of various external agencies. It was established that heating in vacuum, which produces very considerable changes in the density of fast surface states, alters only very slightly the total surface charge. Since the total surface charge consists of the charge at fast and slow surface states, this means that the changes in the charge in the fast states are compensated by almost equal, but opposite in sign, changes of the charge in the slow states. An analysis of the experimental results led us to the conclusion that this compensation is due to mutual transformations of the fast states into slow ones, and conversely, as a result of adsorption-desorption processes. The nature of these transformations was the same as in the neutralization effect. At the same time, these experimental results indicate that the fast surface states are divided into two groups, approximately equal in density, of donor and acceptor states, and that the donor state levels are distributed mainly in the lower half of the forbidden band at the surface while the acceptor levels are mainly in the upper half.

4. Influence of Crystallographic Orientation and the Nature of Surface Treatment

The results given above refer to germanium samples with their surface oriented parallel to the crystallographic plane (111), subjected to the standard treatment in hydrogen peroxide containing an alkali. It was of obvious interest to find to what extent the relationships obtained above are sensitive to a change in the surface orientation and the nature of the surface treatment.

The preliminary results obtained on these questions showed that the processes of activation of recombination and capture centers by heating in vacuum are approximately the same for all surface orientations. The observed differences were only in the form of the recombination curves. Thus, in the case of the (100) orientation, the half-width of the recombination curve (the width at half the amplitude) was only $(5-6)kT/q$ at 300°K, while for the (111) orientation it was $8kT/q$. Since the positions of the maxima of these recombination curves were the same, according to the conventional theory of recombination this means that the energy positions of the recombination levels were different but the values of the ratio of the effective capture cross sections were the same. In the case of the (110) orientation the recombination curves had several characteristic features: they were strongly asymmetric, the positions of their maxima were displaced, and the temperature dependence of the position of the maximum differed considerably from the corresponding dependence for the (111) orientation.

----

[*] This volume, pages 45, 51.

The influence of the chemical composition of the etchant on the recombination properties of a germanium surface was limited to a considerably higher hydrophily of the surface treated in hydrogen peroxide compared with the surface treated in etchants containing hydrofluoric acid. This circumstance is responsible for the considerably lower values of the surface recombination velocity obtained immediately after etching in hydrogen peroxide. Typically after such etching the recombination velocities were 20-30 cm/sec, while after etching in mixtures containing hydrofluoric acid they were usually 200-300 cm/sec. However, the drying of the germanium surface (for example, by pumping in vacuum) left the value of the recombination velocity of a surface etched in hydrofluoric acid practically unaffected, while the drying of a surface etched in hydrogen peroxide increased the surface recombination velocity to the previously mentioned values of 200-300 cm/sec.

5. Impurity Photoconductivity Due to Surface Levels

We conducted experiments aimed to obtain directly the energy topography of the surface levels from the spectral distribution of the photoconductivity in the infrared [12]. The main difficulty here was the separation of the volume and surface effects. We avoided this by using the effect just referred to of the steep rise in the density of surface centers after heating in vacuum at temperatures which produce no volume structure defects. Consequently all the changes in the photoconductivity produced by such heating or by the action of ozonized oxygen (Fig. 5) should be due to a change in the density of surface recombination and capture centers. In this way the existence of two discrete levels, lying approximately 0.04 eV below and 0.1 eV above the middle of the forbidden band, was confirmed. These measurements showed also that there is a system of levels with a continuous energy distribution in the forbidden band, with densities increasing away from the middle of this band. Finally, it was shown that the densities of both discrete and continuous levels vary under the action of heating in vacuum and of ozonized oxygen, in general agreement with the data on recombination and capture.

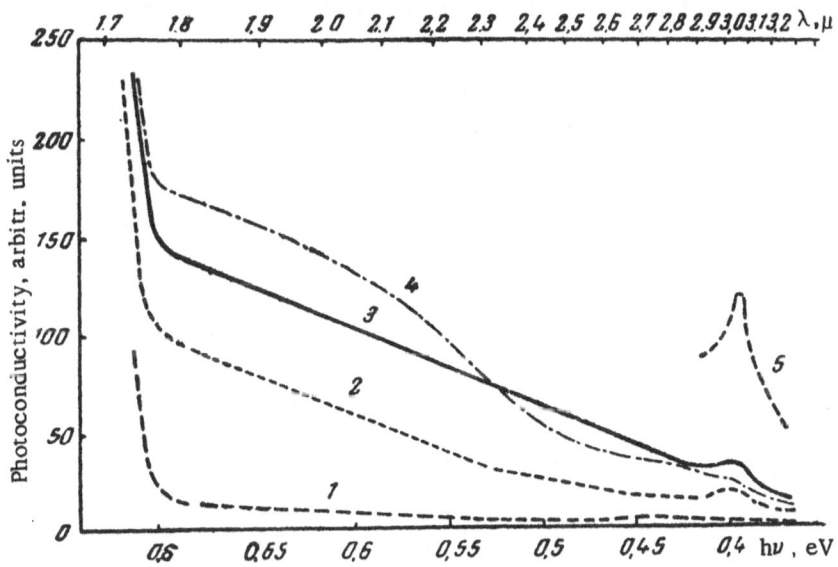

Fig. 5. Spectral distribution of the photoconductivity in 10^{-4} mm Hg vacuum at about 100°K: 1) without preliminary heating of the sample; 2), 3), and 4) after preliminary heating to 400, 500, and 600°K, respectively; 5) another sample subjected first to the action of ozonized oxygen.

Summarizing the analysis of the experimental data, we can say that they indicate complexity of the system of recombination and capture centers on the real surface of germanium. From the available data it follows that the acceptor and donor centers, with discrete levels or with continuously distributed levels, make comparable contributions to the surface recombination capture. The ratios of the separate contributions to the surface recombination capture vary considerably with the amounts of oxygen atoms and water molecules adsorbed on the surface, with temperature and with the crystallographic orientation on the surface. The surface recombination and capture centers are complex structures in which the chief components are oxygen atoms or ions bound at irregular lattice sites near the boundary between the germanium and the layer of surface oxides.

6. Surface Recombination Theory

The general conclusion is that the current theory of surface recombination cannot even qualitatively explain all the available experimental data.

Consequently, we attempted to develop a theory of recombination which allows for the duration of localization of captured carriers at a recombination center. The main physical assumptions of the calculations [13] are as follows.

1. Electrons from the conduction band and holes from the valence band are captured only by the excited levels of a recombination center.

2. The energy positions of the excited levels of a center are related to the energies of the Coulomb interaction between a charged center and carriers of one type, and of the polarization interaction between a neutral center and carriers of the other type.

This assumption, in agreement with Lax's calculations [14], explains the large values of the cross sections for carrier capture of the excited levels of a recombination center.

3. The final stages completing the process of recombination are internal transitions of the captured carriers from the excited levels to the ground level of the recombination center. The probabilities of these transitions in the case when they are not accompanied by radiation should be very strongly dependent on temperature.

A theory of recombination based on these assumptions differs from the usual theory [15] only by the interpretation of the concepts of effective cross sections for the carrier capture of a recombination center. However, in cases when the carrier lifetimes at the excited levels of a center are not negligible compared with the average times for liberation or capture of a charge carrier of sign opposite to that of the center, this interpretation implies additional terms in the relationship for the recombination velocity.

The new relationship for the surface recombination velocity [16] has the following form:

$$s = \frac{N_t \cdot (C_p \cdot C_n)^{1/2} \cdot \frac{n_0 + p_0}{2n_i}}{\cosh \frac{q(\varphi_s - \zeta)}{kT} + \cosh \left(\frac{\varepsilon_t - q\zeta}{kT} \right) + M \cdot \left[\cosh \frac{q(\varphi_s - \eta)}{kT} + \cosh \left(\frac{\varepsilon_t - q\eta}{kT} \right) \right]}$$

which differs from the usual expression by the additional terms in the denominator having M as a multiplier.

Here, as usual, N_t is the density of recombination centers; ε_t is the energy position of a level with respect to the middle of the forbidden band; $q\varphi_s/kT$ is the nondimensional surface potential; C_p, C_n are the cross sections for carrier capture by the excited levels of a recombination center. For convenience we introduce a quantity $q\zeta/kT = \frac{1}{2} \ln (C_p/C_n)$.

The new quantities are given by

$$\frac{q\eta}{kT} = \frac{1}{2} \ln \left(\frac{r_p}{r_n} \cdot \frac{n_1^*}{p_1^{**}} \right) \text{ and } M = \left(\frac{C_p \cdot C_n}{r_p \cdot r_n} \cdot n_1^* \cdot p_1^{**} \right)^{1/2},$$

where r_p and r_n are the probabilities of internal transitions of captured carriers from the excited levels to the ground level of the recombination center; n_1^* and p_1^{**} are the densities of carriers in the bands, which correspond to the coincidence of the Fermi level with the excited levels of the recombination center.

The appearance of the additional terms in the expression for the surface recombination velocity complicates the interpretation of the recombination curve: the position of its maximum and its half-width are related in a much more complex way to the ratio of the capture cross sections of the excited levels and to the energy position of the recombination level. Moreover, the relationship involves new theoretical parameters which are the probabilities of internal transitions. Figure 6 shows a set of recombination curves for the same values of the parameters ε_t, $q\zeta/q\eta$, $q\eta/kT$ but with M varying from zero to infinity, which produces considerable changes in the nature of the recombination curves. Thus the interpretation of the experimental curves becomes more complicated and the number of parameters which need to be determined increases. However, in principle, in every case all the theoretical parameters can be determined experimentally by additional measurements, for example by measuring the relaxation times of the photoconductivity in the infrared.

The more complicated theory gives a unified explanation of all the known experimental relationships governing the surface recombination, including those which could not be explained at all by the conventional theory. Among the latter, we can mention the exponential temperature dependence of the effective capture cross sections and the related complex temperature dependence of the positions of the recombination curve maxima, the insensitivity of the maximum values of the surface recombination velocity to changes in the values of the carrier capture cross sections of the excited levels of a recombination center, and the differences in the form of the recombination curves corresponding to different crystallographic orientations of the germanium surface.

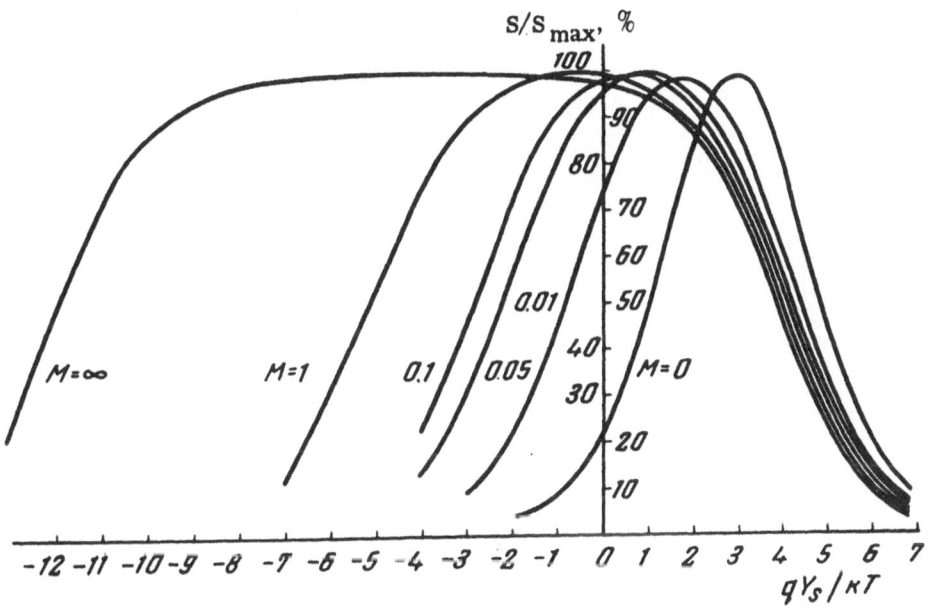

Fig. 6. Set of theoretical recombination curves for fixed values of the parameters ε_t = 4kT, $q\xi$ = 3kT, $q\eta$ = -4kT and M varying from zero to infinity.

In conclusion one must emphasize that the experiments demonstrate conclusively that the generally accepted methods of the surface treatment of germanium and germanium devices are irrational. The oxidation-reduction etching of germanium produces an unstable layer consisting of germanium oxide whose properties change considerably when the ambient conditions are altered. The protection and stabilization of such a layer is a very difficult process and as a rule it impairs the recombination properties of the surface. Because of this it would be very useful to develop essentially different methods for etching germanium so that layers of sulfides, nitrides, carbides and other stable compounds of germanium are formed on the surface.

LITERATURE CITED

1. A. V. Rzhanov, Yu. F. Novototskii-Vlasov, and I. G. Neizvestnyi, Zhur. Tekh. Fiz. $\underline{27}$, 2440 (1957).
2. A. V. Rzhanov, N. M. Pavlov, and M. A. Selezneva, Zhur. Tekh. Fiz. $\underline{28}$, 2645, (1958).
3. A. V. Rzhanov, Yu. F. Novototskii-Vlasov, and I. G. Neizvestnyi, Fiz. Tverd. Tela $\underline{1}$, 1471 (1959).
4. A. V. Rzhanov, Fiz. Tverd. Tela $\underline{2}$, 2433 (1960).
5. Yu. F. Novototskii-Vlasov and I. G. Neizvestnyi, Pribory i Tekh. Eksp. No. 4, 127 (1961).
6. J. R. Schrieffer, Phys. Rev. $\underline{97}$, 641 (1955).
7. R. F. Greene, R. D. Frankl, and J. N. Zemel, Phys. Rev. $\underline{118}$, 967 (1960).
8. A. V. Rzhanov and I. A. Arkhipova, Fiz. Tverd. Tela $\underline{3}$, 1954 (1961).
9. A. V. Rzhanov, N. M. Pavlov, and M. A. Selezneva, Fiz. Tverd. Tela $\underline{3}$, 832 (1961).
10. A. V. Rzhanov, Yu. F. Novototskii-Vlasov, I. G. Neizvestnyi, S. V. Pokrovskaya, and T. P. Galkina, Fiz. Tverd. Tela $\underline{3}$, 822 (1961).
11. A. V. Rzhanov, Fiz. Tverd. Tela $\underline{3}$, 1718 (1961).
12. A. V. Rzhanov and A. F. Plotnikov, Fiz. Tverd. Tela $\underline{3}$, 1557 (1961).
13. A. V. Rzhanov, Fiz. Tverd. Tela $\underline{3}$, 3691 (1961).
14. M. Lax, Phys. Rev. $\underline{119}$, 1502 (1960).
15. W. Shockley and W. Read, Phys. Rev. $\underline{87}$, 835 (1952).
16. A. V. Rzhanov, Fiz. Tverd. Tela $\underline{3}$, 3698 (1961).

THEORY OF THE PHOTOADSORPTION EFFECT
IN SEMICONDUCTORS

F. F. Vol'kenshtein and I. V. Karpenko

Physical Chemistry Institute, Academy of Sciences, USSR;
M. V. Lomonosov State University, Moscow

The photoadsorption effect is the change in the specific adsorption of a surface under the action of illumination. By the specific adsorption of a surface for a particular type of molecule, we mean the number of molecules of this type captured by a unit area of the surface at a given pressure and temperature under conditions of equilibrium with the gaseous phase.

The change in the specific adsorption of a semiconductor surface under the action of illumination can now be regarded as an experimentally established fact (for review of the experimental data see [1] and Sec. 10b in [2]). Usually the change in the specific adsorption is detected from the change of pressure in the adsorption chamber on commencement of illumination. A pressure drop indicates an increase of the specific adsorption; a pressure rise, its reduction. In some cases, illumination enhances the specific adsorption, in others depresses it. Thus we can distinguish the positive and negative photoadsorption effects.

The available experimental data on the photoadsorption effect seem, at first sight, to be somewhat contradictory. They all, however, yield one general conclusion: the sign of the effect depends on the system (the nature of the adsorbent and adsorbate), on the experimental conditions (pressure and temperature), and on the whole previous history of the sample subjected to illumination. Thus, by changing the experimental conditions (for example, by changing the pressure in the gaseous environment) or by varying the method of preparing the sample (for example, by introducing some impurities), we can alter the sign of the photoadsorption effect.

An attempt to develop a theory of the photoadsorption effect is discussed in [3] and [4] (see also Sec. 10a in [2]). The present work is a continuation of these papers. Our aim was to establish a criterion which governs the sign of the effect. The problem is to determine the sign of the photoadsorption effect from the values of the parameters which characterize the nature of the adsorbent and adsorbate, the state of the system as a whole and the experimental conditions.

Let us consider a quantity γ defined by the following expression:

$$\gamma = \frac{\Delta n_s}{n_{0s}} \cdot \frac{p_{0s}}{\Delta p_s} \exp\left(\frac{\varepsilon - V_{0s} + v}{kT}\right) - 1, \tag{1}$$

where n_{0s} and p_{0s} are, respectively, the densities of free electrons and holes in the plane of the semiconductor surface in the absence of illumination, and Δn_s, Δp_s are the increments in the density due to illumination. The meaning of the quantities ε, V_{0s} and v, which occur in Eq. (1), is clear from Fig. 1, which gives the energy-band structure for a semiconductor with a charged surface. The quantity V_{0s} represents the surface-band curvature: for negative charge on the surface the bands are bent upwards, as shown in the figure, i.e., $V_{0s} > 0$; when the surface is charged positively the bands are bent downwards, i.e., $V_{0s} < 0$. The quantity ε represents the position of the Fermi level, denoted by F-F, in the unilluminated sample, reckoned from the middle of the energy gap between the conduction and valence bands. A in Fig. 1 denotes a local surface level representing an adsorbed particle which may be an acceptor or a donor.

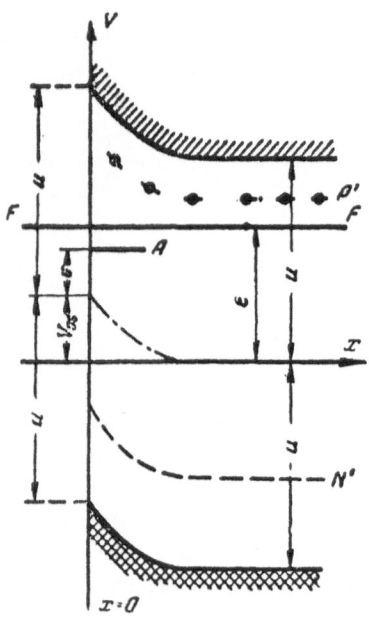

Fig. 1. Energy band model of a semiconductor with charged surface.

In [1, 2] it was shown that the sign of the photoadsorption effect is determined by the sign of γ. In the case of acceptor particles the sign of the effect is the same as that of γ; in the case of donor particles it is opposite to the sign of γ.

The derivation of Eq. (1) is as follows. Some of the levels A (Fig. 1) are populated with electrons, (if these levels are acceptors) or with holes (if the levels are donors), while other levels remain vacant. In other words, some of the particles adsorbed by a unit area of the surface (we shall denote the total number by N) are in a charged state (the number of such particles is N*), while others are neutral (N⁰). Obviously

$$N = N^0 + N^* = N^0 (1 + \eta),$$

where $\eta = N^* / N^0$.

The electron theory of chemisorption [2] shows that equilibrium with the gaseous phase is due only to the neutral particles because the charged particles on the surface do not take part in an exchange with the gaseous environment (provided the electron and hole gas at the surface of the semiconductor can be regarded as nondegenerate). Thus the quantity N⁰ is determined completely by fixing temperature and pressure and it is insensitive to illumination, while the quantity η, representing the degree of population of the local levels A with electrons or holes, changes on illumination. Depending on whether η is increased or decreased by illumination, we have an increase or a decrease of the specific adsorption.

In Eq. (1) the dependence of the sign of the effect on directly measurable parameters is not clear. In fact the density increments Δn_s and Δp_s, which depend on the intensity of the illumination and, for a fixed intensity, also on ε and V_{0s}, are not known. The purpose of the present work is to interpret Eq. (1) for γ, i.e., to calculate Δn_s and Δp_s.

1. Statement of the Problem

We shall consider the general case of a semiconductor containing both acceptor and donor impurities, the concentrations of which are denoted by N' and P' (Fig. 1). We shall assume that the half-space $x \geq 0$ is occupied by the semiconductor, and the half-space $x < 0$ represents the gaseous phase. We shall assume also that the surface of the semiconductor, $x = 0$, is illuminated along the positive direction of x and that this illumination transfers electrons from the valence to the conduction band. Let $n_0(x)$ and $p_0(x)$ be the densities of free electrons and holes in a plane x (where $x \geq 0$) in the absence of illumination, and $\Delta n(x)$ and $\Delta p(x)$ be the excess density increments due to illumination. We shall introduce the notation:

$$n_0 (0) = n_{0s}; \qquad p_0 (0) = p_{0s};$$
$$\Delta n (0) = \Delta n_s; \qquad \Delta p (0) = \Delta p_s,$$

and obviously

$$\Delta n (\infty) = 0, \qquad \Delta p (\infty) = 0.$$

Let $E_0(x)$ and $V_0(x)$ be, respectively, the electric field intensity and the potential energy of an electron in the plane x in the absence of illumination. Clearly $eE_0 = dV_0/dx$, where e is the absolute value of the electron charge. $\Delta E(x)$ and $\Delta V(x)$ denote the corresponding increments due to illumination. We shall use the notation:

$$E_0(0) = E_{0s}; \qquad V_0(0) = V_{0s};$$
$$\Delta E(0) = \Delta E_s; \qquad \Delta V(0) = \Delta V_s,$$

and we shall assume that

$$E_0(\infty) = 0; \qquad V_0(\infty) = 0;$$
$$\Delta E(\infty) = 0; \qquad \Delta V(\infty) = 0.$$

The quantities $\Delta n(x)$ and $\Delta p(x)$, which are of interest to us can be found by the simultaneous solution of Poisson's equation and the equations of continuity of electrons and holes. We shall solve this system of equations on the assumption that

$$\Delta n(x), \ \Delta p(x) \ll p_0(x), \quad \text{if} \quad p_0(x) \gg n_0(x),$$
$$\Delta n(x), \ \Delta p(x) \ll n_0(x), \quad \text{if} \quad n_0(x) \gg p_0(x)$$

and that at all values of x the following condition is obeyed

$$|\Delta V(x)| \ll kT.$$

With these assumptions the equations of continuity have the form

$$\frac{dj_n}{dx} = \frac{dj_p}{dx} = g - \frac{\Delta n}{\tau_n} - \frac{\Delta p}{\tau_p}, \tag{2}$$

where j_n, j_p are the electron and hole currents, respectively ($j_n = j_p$); g is the number of electron–hole pairs excited by illumination in unit volume per unit time; τ_n and τ_p are the lifetimes of electrons and holes. We have

$$\left. \begin{array}{l} j_n = -D_n \left(\dfrac{d\Delta n}{dx} + \Delta n \dfrac{eE_0}{kT} + n_0 \dfrac{e\Delta E}{kT} \right) \\[2mm] j_p = -D_p \left(\dfrac{d\Delta p}{dx} - \Delta p \dfrac{eE_0}{kT} - p_0 \dfrac{e\Delta E}{kT} \right) \end{array} \right\} \tag{3}$$

$$g = \eta \varkappa I_0 \exp(-\varkappa x), \tag{4}$$

where D_n, and D_p are the diffusion coefficients of electrons and holes; I_0 is the intensity of illumination (the number of quanta incident in 1 sec on 1 cm^2) in the x = 0 plane; $\varkappa$ is the adsorption coefficient; η is the quantum yield (we shall assume that $\eta = 1$).

Poisson's equation can be written in the following form:

$$\frac{d\Delta E}{dx} = \frac{4\pi e}{\chi} [(1 + \delta_p) \Delta p - (1 + \delta_n) \Delta n], \tag{5}$$

where χ is the permittivity; δ_p, δ_n are the increments due to the presence in the semiconductor of donor and acceptor impurities, respectively. We note that when P' = 0 (the case of a p-type semiconductor), we have $\delta_p = 0$; if the semiconductor is degenerate (practically complete ionization of the impurity) then $\delta_n \ll 1$. In the case N' = 0 (n-type semiconductor) we have $\delta_n = 0$; then for a degenerate semiconductor $\delta_p \ll 1$.

Simultaneous solution of Eqs. (2) and (5) in the general case meets with mathematical difficulties. Therefore, we shall obtain this solution making the following approximation about the potential, which is good enough for our purpose. We shall assume that*

$$E_0(x) = \begin{cases} = E_{0s}, & \text{when } 0 \leqslant x \leqslant x_0, \\ = 0, & \text{when } x_0 \leqslant x \leqslant \infty, \end{cases} \tag{6}$$

where $x_0 = -V_{0s}/eE_{0s}$. The solutions obtained separately for the regions $0 \leq x \leq x_0$ and $x_0 \leq x \leq \infty$, are then joined in the plane $x = x_0$ (we note that x_0 is eliminated from the final formulas). The expressions for $\Delta n_s = \Delta n(0)$ and $\Delta p_s = \Delta p(0)$ obtained in this way are then substituted into Eq. (1) which gives us the criteria for the positive and negative photoadsorption effects.

2. Calculation of Δn_s and Δp_s

Let us return first to the solution of the continuity equations in the region $0 \leq x \leq x_0$. In this region we shall neglect, as usual, the recombination terms $\Delta n/\tau_n$ and $\Delta p/\tau_p$. Moreover, we shall assume that for all x (in the region $0 \leq x \leq x_0$) we have

$$\left| \frac{\Delta E}{E_{0s}} \right| \leqslant \frac{\Delta n}{n_0}, \quad \frac{\Delta p}{p_0}.$$

The above conditions mean that the field near the surface of the semiconductor is sufficiently strong. Under these conditions it follows from Eqs. (3) and (6) that

$$\left. \begin{aligned} j_n &= -D_n \left(\frac{d\Delta n}{dx} + \Delta n \frac{eE_{0s}}{kT} \right) \\ j_p &= -D_p \left(\frac{d\Delta p}{dx} - \Delta p \frac{eE_{0s}}{kT} \right) \end{aligned} \right\}, \tag{7}$$

and in accordance with Eq. (4), Eq. (2) becomes

$$\left. \begin{aligned} D_n \frac{d}{dx} \left(\frac{d\Delta n}{dx} + \Delta n \frac{eE_{0s}}{kT} \right) &= -\varkappa I_0 \exp(-\varkappa x) \\ D_p \frac{d}{dx} \left(\frac{d\Delta p}{dx} - \Delta p \frac{eE_{0s}}{kT} \right) &= -\varkappa I_0 \exp(-\varkappa x) \end{aligned} \right\}. \tag{8}$$

First integration of the equations (8) gives:

$$\frac{d\Delta n}{dx} + \Delta n \frac{eE_{0s}}{kT} = -\frac{1}{D_n} \{ I_0 [1 - \exp(-\varkappa x)] + j_s \},$$

$$\frac{d\Delta p}{dx} - \Delta p \frac{eE_{0s}}{kT} = -\frac{1}{D_p} \{ I_0 [1 - \exp(-\varkappa x)] + j_s \},$$

where

$$j_s = j_n(0) = j_p(0)$$

(j_s is proportional to the surface recombination velocity).

---

* G. L. Bir [5] used a similar approximation for the solution of the equations of continuity.

Second integration results in

$$\Delta n(x) = A_1 \exp\left[-V_0(x)/kT\right] + \frac{I_0 \exp(-\varkappa x)}{D_n(eE_{0s}/kT - \varkappa)} - \frac{kT(I_0 + j_s)}{eE_{0s} \cdot D_n}$$

$$\Delta p(x) = A_2 \exp\left[V_0(x)/kT\right] - \frac{I_0 \exp(-\varkappa x)}{D_p(eE_{0s}/kT + \varkappa)} + \frac{kT(I_0 + j_s)}{eE_{0s} \cdot D_p}$$

$$\tag{9}$$

where A_1 and A_2 are integration constants which have to be determined.

We shall now return to the region $x_0 \leq x \leq \infty$. In this region, according to Eqs. (3) and (6), we have

$$j_n = -D_n\left(\frac{d\Delta n}{dx} + n_0 \frac{e\Delta E}{kT}\right)$$

$$j_p = -D_p\left(\frac{d\Delta p}{dx} - p_0 \frac{e\Delta E}{kT}\right)$$

$$\tag{10}$$

We shall neglect the term g in Eq. (2), which is permissible if $\varkappa$ is not too small. On this assumption, and taking into account that in the region considered n_0 = const. and p_0 = const., the equations of continuity (2) can be written thus:

$$\frac{d^2\Delta n}{dx^2} + \frac{en_0}{kT} \cdot \frac{d\Delta E}{dx} = \frac{1}{D_n}\left(\frac{\Delta n}{\tau_n} + \frac{\Delta p}{\tau_p}\right)$$

$$\frac{d^2\Delta p}{dx^2} - \frac{ep_0}{kT} \cdot \frac{d\Delta E}{dx} = \frac{1}{D_p}\left(\frac{\Delta n}{\tau_n} + \frac{\Delta p}{\tau_p}\right)$$

Substituting here Eq. (5), we obtain

$$\frac{d^2\Delta n}{dx^2} - (\alpha_{nn} + \beta_{nn})\,\Delta n = (\alpha_{pn} - \beta_{pn})\,\Delta p$$

$$\frac{d^2\Delta p}{dx^2} - (\alpha_{pp} + \beta_{pp})\,\Delta p = (\alpha_{np} - \beta_{np})\,\Delta n$$

$$\tag{11}$$

where

$$\alpha_{nn} = 1/\tau_n D_n; \quad \beta_{nn} = n_0(1 + \delta_n) \cdot 4\pi e^2/\chi kT$$

$$\alpha_{pp} = 1/\tau_p D_p; \quad \beta_{pp} = p_0(1 + \delta_p) \cdot 4\pi e^2/\chi kT$$

$$\alpha_{np} = 1/\tau_n D_p; \quad \beta_{np} = p_0(1 + \delta_n) \cdot 4\pi e^2/\chi kT$$

$$\alpha_{pn} = 1/\tau_p D_n; \quad \beta_{pn} = n_0(1 + \delta_p) \cdot 4\pi e^2/\chi kT$$

$$\tag{12}$$

Solutions of the system (11) have the form

$$\Delta n(x) = B_1 \sqrt{\frac{\alpha_{pn} - \beta_{pn}}{\alpha_{pp} - \beta_{nn}}}\, e^{-k_1 x} + B_2 \sqrt{\frac{\alpha_{pn} - \beta_{pn}}{\alpha_{nn} - \beta_{pp}}}\, e^{-k_2 x}$$

$$\Delta p(x) = B_1 \sqrt{\frac{\alpha_{np} - \beta_{np}}{\alpha_{nn} - \beta_{pp}}}\, e^{-k_1 x} + B_2 \sqrt{\frac{\alpha_{np} - \beta_{np}}{\alpha_{pp} - \beta_{nn}}}\, e^{-k_2 x}$$

$$\tag{13}$$

where

$$k_1 = \sqrt{\alpha_{nn} + \alpha_{pp}};$$

$$k_2 = \sqrt{\beta_{nn} + \beta_{pp}},$$

and B_1, B_2 are integration constants.

The unknown constants A_1, A_2, and B_1, B_2 in Eqs. (9) and (13) can be determined from the conditions which ensure that the solutions are continuous in the plane $x = x_0$. These conditions are written thus:

$$j_n^{\mathrm{I}}(x_0) = j_n^{\mathrm{II}}(x_0); \quad \Delta n^{\mathrm{I}}(x_0) = \Delta n^{\mathrm{II}}(x_0) \left.\right\}$$
$$j_p^{\mathrm{I}}(x_0) = j_p^{\mathrm{II}}(x_0); \quad \Delta p^{\mathrm{I}}(x_0) = \Delta p^{\mathrm{II}}(x_0) \left.\right\}, \tag{14}$$

where the indices I and II represent, respectively, the regions $0 \leq x \leq x_0$ and $x_0 \leq x \leq \infty$. Substituting the expressions (7) and (9) into the left-hand parts of Eq. (14) and the expressions (10) and (13) into the right-hand parts of the same equation, we obtain the required unknowns A_1 and A_2. In the case of a semiconductor which is degenerate in the region $x \geq x_0$ (we shall consider only this case), A_1 and A_2 assume the following form:

$$A_1 = (\sqrt{\tau_i/D_i} + kT/eE_{0s}D_n)(I_0 + j_s) \left.\right\}$$
$$A_2 = (\sqrt{\tau_i/D_i} - kT/eE_{0s}D_p)(I_0 + j_s) \left.\right\}. \tag{15}$$

Here (and later) we take i = p in the case of a n-type semiconductor and i = n in the case of a p-type semiconductor.

Substituting (15) in (9) and taking x = 0, we obtain

$$\Delta n_s = \left(\sqrt{\frac{\tau_i}{D_i}} + \frac{kT}{eE_{0s}D_n} \right)(I_0 + j_s)\exp(-V_{0s}/kT) - $$
$$ - \frac{kT}{eE_{0s}D_n}\left(\frac{I_0}{1 - eE_{0s}/kT\varkappa} + j_s \right) \left.\right\}$$
$$\Delta p_s = \left(\sqrt{\frac{\tau_i}{D_i}} - \frac{kT}{eE_{0s}D_p} \right)(I_0 + j_s)\exp(V_{0s}/kT) + $$
$$ + \frac{kT}{eE_{0s}D_p}\left(\frac{I_0}{1 + eE_{0s}/kT\varkappa} + j_s \right) \left.\right\}. \tag{16}$$

We note that for a majority carrier density (in the interior of a crystal) of $10^{14}\,\mathrm{cm^{-3}}$ and for D_n, $D_p = 100$ $\mathrm{cm^2/sec}$, $\tau_i = 10^{-4}$ sec, $\chi = 9$, $\varkappa = 10^6\mathrm{cm^{-1}}$, $|V_{0s}| = 0.1$ eV, kT = 0.04 eV, we have

$$\varkappa \gg \frac{e|E_{0s}|}{kT} \gg \frac{1}{\sqrt{\tau_i D_i}}\exp(|V_{0s}|/kT). \tag{17}$$

In this case the expressions for Δn_s and Δp_s, according to Eq. (16), assume the following final form:

$$\Delta n_s = (I_0 + j_s)\sqrt{\frac{\tau_i}{D_i}}\exp(-V_{0s}/kT) \left.\right\}$$
$$\Delta p_s = (I_0 + j_s)\sqrt{\frac{\tau_i}{D_i}}\exp(+V_{0s}/kT) \left.\right\}. \tag{18}$$

3. Discussion of Results

Let us return to Eq. (1) for γ. Substituting Eq. (18) into Eq. (1) and allowing for the fact that (cf. Fig. 1)

$$n_{0s} = C_n \exp[-(u - \varepsilon + V_{0s})/kT];$$
$$p_{0s} = C_p \exp[-(u + \varepsilon - V_{0s})/kT],$$

we have (assuming that $C_n = C_p$)

$$\gamma = \exp(-\varphi/kT) - 1,$$

where

$$\varphi = \varepsilon + V_{0s} - v. \tag{19}$$

It is clear that the sign of γ and therefore the sign of the photoadsorption effect is governed by the sign of φ. We thus have

a) for acceptor particles
positive effect (photoadsorption) if $\varphi < 0$
negative effect (photodesorption) if $\varphi > 0$

b) for donor particles
positive effect (photoadsorption) if $\varphi > 0$
negative effect (photodesorption) if $\varphi < 0$

$$\tag{20}$$

It is clear from Eqs. (19) and (20) that for a given adsorbent and a given adsorbate (i.e., for fixed v), the sign of the effect is governed by the position of the Fermi level in the interior of the crystal ε and by the degree of surface curvature of the bands V_{0s} (cf. Fig. 1). We should note that ε and V_{0s} depend on the previous history of the sample (i.e., on the nature of the treatment to which it had been subjected) and on temperature; moreover, V_{0s} depends on the extent of the coverage of the surface with adsorbed particles (i.e., on the pressure in the gaseous phase). Thus Eqs. (19) and (20) show the dependence of the sign of the effect on the nature of the adsorbent and adsorbate, on the experimental conditions, and on the history of the sample before illumination.

Figure 2 (compare this with Fig. 1) shows the region of possible values of ε and V_{0s} in which the Boltzmann distribution of free carriers applies to the whole volume of the crystal (the case when the Fermi level does not intersect the energy bands at any point). The thick lines AA and BB divide this region, in accordance with Eqs. (19) and (20), into areas of positive and negative effects suitably distinguished in the figure. The intercept v represents the position, in the energy spectrum, of the local surface levels corresponding to adsorbed particles of the type considered (obviously, depending on the nature of the adsorbent and adsorbate, we can have $v > 0$ and $v < 0$, cf. Fig. 1); the intercept V_{0s}^{*} represents the initial curvature of the bands in the absence of adsorbed particles on the surface, due to the surface states of nonadsorptive origin (obviously, depending on the nature of the preliminary surface treatment, we can have $V_{0s}^{*} > 0$ and $V_{0s}^{*} < 0$).

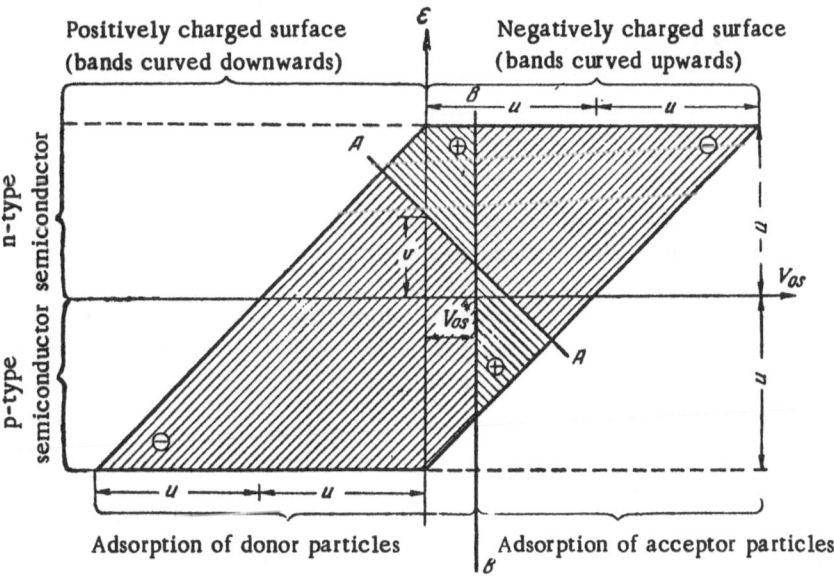

Fig. 2. Regions of positive and negative photoadsorption effects.

It is clear that on displacement of the Fermi level ε (for V_{0s} = const.), or on change of the degree of band curvature V_{0s} (for ε = const.), or when both these parameters are altered simultaneously, the sign of the photoadsorption effect may be reversed.

It follows that we can expect a reversal of the sign of the photoadsorption effect on varying the initial electrical conductivity of the semiconductor (change of ε) as well as on varying of the work function (change of V_{0s}). It would be of interest to check experimentally these theoretical predictions.

From Fig. 2 it also follows that the regions of the positive and negative effects should expand or contract depending on the preliminary treatment of the surface (a change of V_{0s}^*). The vertical line BB in Fig. 2 is displaced by the preliminary treatment to the left or right, parallel to itself.

4. Comparison with Experiment

Let us consider now the available experimental data.

1. Romero-Rossi and Stone [6] investigated the adsorption of oxygen on zinc oxide at room temperature. At low oxygen pressures they found photoadsorption, which, however, was replaced by photodesorption when the pressure was increased.

This was to be expected from Fig. 2. Romero-Rossi and Stone were working in the region above the V_{0s} axis and to the right of the BB vertical in Fig. 2 (this represents an n-type semiconductor, adsorption of acceptor particles). An increase of pressure represents an increase of V_{0s}, i.e., a displacement from left to right in Fig. 2. It follows from Fig. 2 that if at low V_{0s} we are in the region of the positive effect (photoadsorption), then on increase of V_{0s} we move over to the negative-effect region (photodesorption). This was exactly what Romero-Rossi and Stone observed.

2. They also used the same system at 400°C and obtained the opposite result: the photodesorption observed at low pressures was replaced with photoadsorption when the pressure was increased [6].

It should be noted that the adsorption of oxygen on ZnO at 400°C is irreversible. Stone and others [6, 7, 8] explained this by the transformation of the adsorbed O^- ions into O^{2-} ions as a result of the emergence on the surface of interstitial Zn^+ ions and the reaction

$$O_a^- + Zn_i^+ \rightarrow Zn^{2+}O^{2-},$$

where O_a^- represents an adsorbed oxygen ion, and Zn_i^+ is an interstitial zinc ion. This reaction takes place only at sufficiently high temperatures (over 200°C) which ensure diffusion of interstitial Zn^+ ions from the interior of the semiconductor to its surface under the action of an electric field produced by the chemisorbed O^- ions. According to this point of view, shared by many, the adsorption of oxygen on zinc oxide at high temperatures simply represents the growth of the crystal accompanied by a reduction of the excess zinc content.

Such adsorption leads to a lowering of the Fermi level in the interior of the semiconductor (reduction of ε). As the oxygen pressure is increased the surface coverage increases, the number of interstitial Zn^+ ions expelled from the interior to the surface grows, the excess zinc content in the crystal decreases, and ε decreases. In Fig. 2, we then move downwards (remaining inside the region lying above the V_{0s} axis to the right of the BB vertical). We see that if at high ε (low pressures) we are in the negative-effect region (photodesorption), then at sufficiently low ε (high pressures) we can reach the positive-effect region (photoadsorption). This is exactly what Romero-Rossi and Stone found.

3. Barry [7], dealing with the same system (oxygen on zinc oxide), investigated the influence of the preliminary treatment of the sample on the sign of the photoadsorption effect. A sample was heated at high temperatures in an oxygen atmosphere and then cooled to room temperature, at which adsorption took place. On untreated samples, photodesorption was observed; while on samples treated as above, photoadsorption was found.

Here the sample treatment lowered the Fermi level ε (the mechanism was considered in the preceding section). According to Fig. 2, this means, as seen above, a transition from the negative to the positive region, which was actually observed.

Barry's results agree with those of Kwan and Fujita [9] and of Terenin and Solonitzin [10]. The latter investigated the photoadsorption effect of oxygen on zinc oxide samples of various degrees of departure from stoichiometry. It was found that the oxidation of a sample was always accompanied by a lowering of the Fermi level (reduction of ε). Consequently, if photodesorption occurred on reduced samples (with a high concentration of excess zinc), then on oxidized samples (with a lower concentration of excess zinc) this could be replaced by photoadsorption. Precisely this result was observed [9, 10].

4. Exactly the opposite was obtained by Romero-Rossi and Stone [6], according to whom photodesorption is observed on ZnO samples with a lower content of excess zinc and photoadsorption in those with a higher content.

The contradiction between the experimental results of different workers can be understood if we allow for the fact that the sample treatment may (and as a rule does) alter simultaneously both ε and V_{0s}. Thus in the case of zinc oxide containing excess zinc in the form of interstitial Zn^+ ions (the concentration of which we shall denote by N_i), both ε and V_{0s} are functions of N_i. It can be easily shown that

$$\left.\begin{array}{l} \varepsilon = u + kT \ln (N_i / C_n); \\ V_{0s} = \dfrac{2\pi e^2}{\chi} \cdot \dfrac{\sigma^2}{N_i}. \end{array}\right\} \tag{21}$$

Here, σ is the surface charge expressed in units of electron charge. Substituting Eq. (21) in Eq. (19) we obtain

$$\varphi = (u - v) + kT \ln \frac{N_i}{C_n} + \frac{2\pi e^2}{\chi} \cdot \frac{\sigma^2}{N_i}. \tag{22}$$

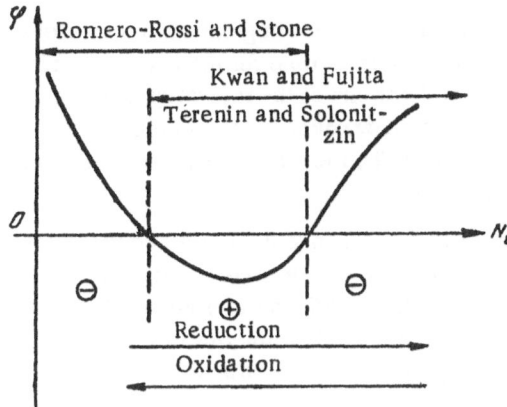

Fig. 3. Dependence of the sign of the photoadsorption effect on the excess zinc content in ZnO.

The dependence of φ on N_i at fixed σ is shown schematically in Fig. 3, in accordance with Eq. (22). The vertical dashed lines in Fig. 3 separate the regions of positive and negative photoadsorption effects.

We see that as a sample is gradually oxidized (i.e., as N_i is reduced) the function $\varphi(N_i)$ changes its sign twice. The disagreement between the results of Romero-Rossi and Stone on the one hand, and of Kwan and Fujita, Terenin and Solonitzin, on the other, can be understood if we assume that these workers were investigating different regions of N_i values, as shown in Fig. 3.

LITERATURE CITED

1. A. N. Terenin, Problemy Kinetiki i Kataliza 8, 17 (1955).
2. F. F. Vol'kenshtein, Electron Theory of Catalysis on Semiconductors (Fizmatgiz, 1960); Theorie electronique de la catalyse sur les semiconducteurs (Edition Masson, Paris, 1960).
3. F. F. Vol'kenshtein and Sh. M. Kogan, Izvest Akad. Nauk SSSR, Otdel.Khim.Nauk 1536 (1959); J. Chim. phys. 55, 483 (1958).
4. F. F. Vol'kenshtein, Kinetika i Kataliz 2, 481 (1961); Trans. Faraday Soc. 31, 209 (1961).
5. G. L. Bir, Fiz. Tverd. Tela 1, 67 (1959).
6. F. Romero-Rossi and F. S. Stone, Second International Congress on Catalysis, Paris, 1960, Sec. II, Preprint 72.
7. T. I. Barry, ibid., Preprint 70.
8. T. I. Barry and F. S. Stone, Proc. Roy. Soc. (London) A255, 124 (1960).
9. Y. Fujita and T. Kwan, Bull. Chem. Soc. Japan 31, 379 (1958).
10. A. N. Terenin and Y. P. Solonitzin, Discussions Faraday Soc. 28, 28 (1959).

KINETICS OF SURFACE PHENOMENA IN SEMICONDUCTORS IN THE CASE OF LARGE CHANGES OF THE SURFACE POTENTIAL

A. É. Yunovich

M. V. Lomonosov State University, Moscow

As found by several workers [1-4], the investigation of the field-effect kinetics in germani m can yield information on the energy positions of the surface states and the effective cross sections for the capture of electrons and holes. Calculations of the frequency dependence of the field effect and of the relaxation time of transient processes, both in the case of carriers of one sign and in that of carriers of both signs, were carried out in the approximation which assumes a weak perturbation of the equilibrium [5-7], i.e., it was assumed that an external electric field changes the surface potential by much less than kT/q.

The relaxation of the field effect in a strongly depleted layer, when the change in the surface potential is large but only the liberation of majority carriers from the surface states is important, was considered by Rupprecht [8]. In this case, the relaxation time is independent of the density of surface states and of the energy band curvature at the surface. Rupprecht investigated experimentally the field-effect kinetics in silicon and calculated, using his own analysis, the effective cross sections and energy levels of two surface capture centers.

Yunovich and Tikhonov [15] examined the field effect in silicon and found that its relaxation time τ depends appreciably on the magnitude and sign of the induced charge. When the surface layer was depleted of charge, the value of τ varied linearly with the applied field. Moreover, the experiment showed that in some cases the role of minority carriers in the field effect in silicon is negligible.

The nonexponential decay of the excess conductivity with time and the dependence of the relaxation processes on the magnitude and sign of the charge were investigated also by Snitko and Litovchenko [3, 4]. Similar phenomena were studied by Morrison in germanium and silicon at the temperatures of liquid nitrogen and "dry ice" [9]. Morrison associated the characteristic features of the relaxation processes in the field effect with minority carriers but he did not reject other possible mechanisms.

Lindley and Bunbary [10] investigated the relaxation of the field effect in germanium at low temperatures and found that majority carriers played the dominant role in relaxation processes.

In the present work, we consider the kinetics of electron exchange between majority carriers and surface states when the excess charge in the space-charge layer at the surface is arbitrary in magnitude. The results of these calculations agree with the experimental data presented in the work just cited.

The dependence of the relaxation time of the field effect on the magnitude of the excess charge can obviously be used to determine the surface potential of a semiconductor by a new method. This method does not require a knowledge of the minimum value of the space-charge layer conductivity or of the weak screening of the external field by surface levels, in contrast to the widely used method described by Garrett, Brattain, Brown and Rzhanov [11-13].

Calculations show that an investigation of the dependence of the field-effect relaxation on the applied field and temperature should make it possible to determine the density and energies of surface states as well as the effective cross section for the capture of majority carriers.

1. Statement of the Problem

We shall assume that the semiconductor is impurity p-type and that in its interior and on the surface the minority carriers (electrons) can be neglected. According to the theory of the semiconductor surface [11], and with the appropriate assumptions,* the expressions for the hole excess Γ_p in the space charge layer and the hole density at the surface p_s can be written in the form

$$\Gamma_p = \pm p_0 L_D \sqrt{e^{-y_s} - 1 + y_s}, \quad p_s = p_0 e^{-y_s}, \tag{1}$$

where $L_D = (2\pi q \beta p_0 / \varepsilon)^{-1/2}$ is the Debye length; p_0 is the volume hole density; $y_s = \beta(\psi_s - \psi_0)$ is the dimensionless band curvature at the surface; $\beta = q/kT$; ψ_s and ψ_0 are the potentials at the surface and in the interior of the semiconductor, respectively. The plus sign corresponds to negative y_s and the minus sign to positive y_s.

If n_m electrons are induced on the metal plate of the metal—semiconductor capacitor, the change in the hole excess $\Delta \Gamma_p = \Gamma_p - \Gamma_{p0}$ produced by the external field is given by

$$\Delta \Gamma_p = n_m - \sum_t \Delta p_t, \tag{2}$$

where $\Delta p_t = p_t - p_{t_0}$ is the change in the number of holes at the t-th surface level. The rate of change in the number of holes at the t-th level can be written, assuming the independence of the degrees of level population, in the form

$$\frac{d\Delta p_t}{dt} = + \alpha_{pt} \left[n_t p_s - p_{s0} \frac{f_{t0}}{1 - f_{t0}} p_t \right], \tag{3}$$

where α_{pt} is the hole-capture coefficient; $n_t = N_t f_t$ and $p_t = N_t(1 - f_t)$ are, respectively, the electron hole densities in the traps; N_t and f_t are, respectively the density and the Fermi function of the states considered. Let $\Delta p_s = p_s - p_{s0}$ denote the change in the hole density at the surface (the subscript zero refers to equilibrium conditions). We note that the quantities p_s and Δp_s are known single-valued functions of the quantity $\Delta \Gamma_p$ on the basis of the equalities in Eq. (1).

If we assume that one surface level plays the dominant role, it follows then from Eqs. (2) and (3) that

$$\frac{d\Delta \Gamma_p}{dt} = - \alpha_{pt} \Delta \Gamma_p \left(p_{s1} + p_s + n_{t0} \frac{\Delta p_s}{\Delta \Gamma_p} \right) + \frac{dn_m}{dt} + n_m \alpha_{pt} (p_{s1} + p_s). \tag{4}$$

We are using here the Shockley notation

$$p_{s1} = p_{s0} f_{t0} / (1 - f_{t0})$$

for the hole density at the surface when the Fermi level passes through the trap level Analysis of the above equation allows us to derive quantitative relationships describing the process of the field-effect relaxation. The assumption that there is only one center active in hole capture at the surface is the well-known limitation of the theory. However, in the case of a high density of surface states with a single energy level, the phenomena discussed should proceed in agreement with this assumption. Appendix 1 shows which conclusions remain valid in the case of several surface levels.

---

* Under nonequilibrium conditions the expression (1) is valid if the space-charge disperses sufficiently rapidly, i.e., when the characteristic times are large compared with the Maxwellian relaxation time $\tau_M = \varepsilon / 4\pi q \mu_p p$.

2. Relaxation of the Field Effect

Since the right-hand part of Eq. (4) contains a product of the function n_m and an unknown function of time p_s, Eq. (4) is not solved by squaring when the number of the charges induced by the external field depends in an arbitrary fashion on time.

The problem of the re-establishment of equilibrium between the surface states and the volume charge after the instantaneous removal of the field can be analyzed quite simply. Under experimental conditions, the switching off of the external field can be regarded as instantaneous if the capacitance of the metal—semiconductor system is much smaller than the capacitance of the space-charge layer [8]. The switching off can be represented by the conditions

$$t < 0 \quad n_m = n_{m_0}; \quad t \geqslant 0 \quad n_m = 0.$$

Then at t > 0 in Eq. (4), the second and third term on the right vanish and this equation becomes

$$\frac{d\Delta\Gamma_p}{dt} = -\frac{\Delta\Gamma_p}{\tau(\Delta\Gamma_p)}, \tag{5}$$

where

$$\tau = \frac{1}{\alpha_{pt}\left(p_{s1} + p_s + n_{t0}\dfrac{\Delta p_s}{\Delta\Gamma_p}\right)} \tag{5a}$$

is the relaxation time, and the dependence of p_s and Δp_s on $\Delta\Gamma_p$ is given by the parametric formulas of Eq. (1). It must be pointed out that if the equilibrium between the surface states and the interior of the semiconductor was not disturbed by a change of the external field but by some other agency, the re-establishment of the equilibrium when this agency ceases to act is given by Eq. (5). Such an external agency may, for example, be the transfer of holes from the valence band to the surface states upon illuminating the semiconductor with infrared radiation. Moreover, if on illumination with wavelengths in the fundamental absorption band the minority carriers rapidly reach the state of equilibrium with the surface traps, the establishment of equilibrium between the surface states and the normal band is also given by Eq. (5).

Several remarks must be made about the meaning of the quantity τ, the relaxation time. The most precise meaning of this quantity is obtained in the case when $\tau(\Delta\Gamma_p) = $ const., i.e., when the latter is independent of time. Then $\Delta\Gamma_p$ decays exponentially to the equilibrium value and the relaxation time determines the reciprocal of the power of the exponential. If, however, $\tau \neq$ const., then the decay law is nonexponential. In this case, the meaning of the relaxation time is given by Eqs. (5) and (5a), where it represents the dependence of the derivative $d\Delta\Gamma_p/dt$ on the quantity $\Delta\Gamma_p$.

In general the dependence of $\Delta\Gamma_p$ on time is given by the integral

$$\int_{\Delta\Gamma_p(0)}^{\Delta\Gamma_p} \frac{d\Delta\Gamma_p}{\Delta\Gamma_p} \cdot \tau(\Delta\Gamma_p) = -t, \tag{6}$$

which, in general, can be calculated using Eqs. (1) and (5a). The value of $\Delta\Gamma_p$ is determined by the value of n_{m0}, as well as by the density and energy positions of the levels (but it is independent of α_{pt}). The dependence $\Delta\Gamma_p(n_{m0})$ has been considered in detail in the work of Lashkarev [14], Garrett and Brattain [11]. In the case of strong screening of the external field by the surface states, when for t < 0, $|\Delta\Gamma_p| < \Delta p_t$, the switching off of the field alters at first only the number of holes in the space-charge layer, and $\Delta p_t(0) \approx -n_{m0}$. In this case, the initial change in the hole excess is

$$\Delta\Gamma_p(0) = n_{m0}.$$

Let us consider some special cases since the integral of Eq. (6) does not give a clear picture of the relaxation process.

2.1 Depletion of the Space-Charge Layer

When $\Delta \Gamma_p < 0$ and its absolute magnitude is large, so that

$$p_s \ll p_{s1}; \quad p_s \ll p_{s0}; \quad \Delta p_s \cong -p_{s0}. \tag{6a}$$

Eq. (5) becomes

$$\frac{d\Delta \Gamma_p}{dt} = -\Delta \Gamma_p \cdot \alpha_{pt} p_{s1} + \alpha_{pt} n_{t0} p_{s0}, \tag{7}$$

and its general solution is

$$\Delta \Gamma_p = [\Delta \Gamma_p(0) - p_{t0}] e^{-\alpha_{pt} p_{s1} t} + p_{t0}. \tag{7a}$$

In the case of weak population of the surface states with holes or strong depletion of the surface after switching off the external field, i.e., if

$$|\Delta \Gamma_p| \gg p_{t0}, \tag{8}$$

the decay of $\Delta \Gamma_p$ is purely exponential with a relaxation time

$$\tau^-_{\text{large}} = \frac{1}{\alpha_{pt} p_{s1}} . \tag{8a}$$

This case was considered by Rupprecht [8]. The relaxation time in Eq. (8a) is independent of the energy band curvature and the semiconductor surface.

If the surface states are strongly populated:

$$\tag{9}$$

$$|\Delta \Gamma_p| \ll p_{t0},$$

then the hole excess varies linearly with time:

$$\Delta \Gamma_p = \Delta \Gamma_p(0) + \alpha_{pt} n_{t0} p_{s0} t. \tag{9a}$$

This corresponds to a relaxation time (in the sense discussed above)

$$\tau^-_{\text{aver}} = \frac{|\Delta \Gamma_p|}{\alpha_{pt} n_{t0} p_{s0}} , \tag{9b}$$

i.e., the relaxation time is proportional to the excess hole charge (we note that $n_{t0} p_{s0} = p_{t0} p_{s1}$). The conditions (6) and (9) give the lower and upper limits of $\Delta \Gamma_p$. It is possible to satisfy these conditions simultaneously because in the depleted layer $|\Delta \Gamma_p|$ is proportional to the square root of y_s, and the hole density on the surface varies exponentially with the surface potential.

2.2 Weak Disturbance of the Equilibrium

It follows from our considerations that the conditions for a weak disturbance of the equilibrium have the form

$$|\Delta \Gamma| \ll p_0 L_D; \quad |\Delta p_s| \ll p_{s0}. \tag{10}$$

In this case the decay is exponential and the relaxation time is

$$\tau_{\text{small}} = \frac{1 - f_{t0}}{\alpha_{pt} p_{s0}} \cdot \frac{\left(-\dfrac{d\Gamma_p}{dy_s}\right)_{y_{s0}}}{\left(-\dfrac{d\Gamma_p}{dy_s}\right)_{y_{s0}} + n_{t0}(1 - f_{t0})} , \tag{10a}$$

which is the same as in [5, 7]. The case of strong screening of the external field by the surface states is of interest. This happens when the density of these states is high and the following condition is satisfied

$$\left(-\frac{d\Gamma_p}{dy_s}\right)_{v_{s0}} \ll n_{l0}\,(1-f_{l0}).\tag{11}$$

Then

$$\tau_{\text{small}} = \frac{\left(-\dfrac{d\Gamma_p}{dy_s}\right)_{v_{s0}}}{\alpha_{pl}\,n_{l0}\,p_{s0}}.\tag{11a}$$

2.3. Enrichment of the Space-Charge Layer

When $\Delta\Gamma_p > 0$ and its absolute value is large, then

$$p_s \gg p_{s0}.\tag{12}$$

In this case

$$\tau^+ = \frac{1}{\alpha_{pl}\left[p_{s1}+p_s\left(1+\dfrac{n_{l0}}{\Delta\Gamma_p}\right)\right]}\;,\tag{12a}$$

and, since p_s depends on $\Delta\Gamma_p$, it is quite difficult to obtain a simple approximation for the solution of Eq. (5).

We note that if p_{s1} is not much larger than p_{s0}, i.e., if the surface Fermi level at equilibrium is not much higher than the trap level, than for equal $|\Delta\Gamma_p|$, the following inequality applies

$$\tau^+ \ll \tau^-.\tag{13}$$

If, however, $p_{s1} \gg p_{s0}$, then another inequality holds:

$$\tau^+ < \tau^-.$$

Thus, if the number of holes in the space-charge layer is greater than the equilibrium value, the re-establishment of equilibrium is, as a rule, much faster than if the number of holes is less than the equilibrium value. When a field of a given sufficiently high intensity is switched on and off, a difference between the relaxation times of the surface states should be observed.

The special cases just considered permit a simple interpretation of the experimental data. If the dependence of the induced charge n_m on time differs from that given above, the solution of Eq. (4) can be obtained only by numerical methods. In Appendix 2, Eq. (4) is rendered in a nondimensional form and a method of solving it is given for the case when n_m varies exponentially with time.

3. Some Consequences of the Theory

If Eqs. (8) and (8a) are satisfied in the case of strong depletion, and the equalities (9), (9a), and (9b) in the case of moderate depletion, then we can easily determine the hole density in the surface states, p_{t0} from the ratio of the relaxation times

$$p_{l0} = \frac{\tau_{\text{large}}}{\tau_{\text{aver}}^-}\,|\Delta\Gamma_p|_{\text{aver}}\tag{14}$$

All the quantities in the right-hand part of the above equation can be measured experimentally.

If conditions (9), (9a), and (9b) are satisfied for moderate depletion and equalities (11) and (11a) are valid for a small signal, then from the ratio of the relaxation times we can find the derivative of the equilibrium hole excess with respect to the dimensionless surface potential:

$$\left(-\frac{d\Gamma_p}{dy_s}\right) = \frac{\tau_{\text{small}}}{\tau_{\text{aver}}^-}|\Delta\Gamma_p|_{\text{aver}}\tag{15}$$

The quantity $(d\,\Gamma_p/\,dy_s)_{y_{s_0}}$ is a single-valued function of the surface potential ψ_{s_0} [cf. Eq. (1)]. Equation (15) suggests the interesting possibility of experimental determination of the surface potential. This possibility depends on the inequalities (9) and (11) which represent a high density of surface levels and strong screening of the external field by the surface states. Obviously to use this method it is also necessary to satisfy the conditions discussed in the introduction to the present paper: carriers of one sign only and effectively only one surface level.

In the majority of investigations of the surface states in semiconductors, the surface potential has been determined by the large-signal field-effect method [11-13], the use of which assumes relatively weak screening of the external field by the surface states and the important role of carriers of both signs. The interest in the method proposed here for measuring ψ_{s_0} lies in the conditions of its application which are opposite to those for the large-signal field-effect method.

4. Discussion of Results

We shall now compare the results of our calculations with the experimental data reported by several workers.

Experiments on the field-effect kinetics in silicon described by Yunovich and Tikhonov [15] agree with the theoretical predictions. In the work just cited, as well as in [4, 9, 10], a large difference was found between the relaxation times for switching off the fields of different polarity. The switching on and off also produced considerable differences between the relaxation times [cf. the inequality (13)]. On the basis of our model, we can also explain the experimental observation that the relaxation time depends linearly on the induced charge. However, if $\tau \sim |\Delta\Gamma_p|$, the dependence $\Delta\Gamma_p(t)$ should be close to the linear law of Eq. (9a), while the experiments [15] showed only a small departure from the exponential. The reasons for this can be found only by additional experiments. It should be noted that a nonrigorous fulfillment of the theoretical conditions may alter the nearly exponential law of the excess charge variation and the measured effective relaxation time may depend on the initial value $\Delta\Gamma_p(0)$. A nearly linear decay law of the field effect was observed by Morrison [9], Litovchenko and Snitko [4]. Their experimental conditions were such that one could assume that the theoretical conditions were satisfied (carriers of one sign only). Morrison's explanation [9], using a model with the liberation of minority carriers from the surface states, seems less likely.

The dependence of the relaxation time on an additional constant field can be explained on the basis of the approximations considered in the present paper. For example, a positive constant voltage applied to the metal—semiconductor capacitor corresponds to a reduction of the number of holes at the surface p_{s_0} and a reduction of the number of electron-filled traps n_{t_0} (in the case of strong screening of the field by the surface states the change in p_{s_0} is relatively small). This should increase the coefficient in the dependence of τ_{aver} on $|\Delta\Gamma_p|$, given by Eq. (9b), and increase τ_{small} of Eq. (11a). Experiments [15] confirmed this. The value of τ_{large}^- should be independent of the constant field. The dependence of τ on the induced charge, investigated for silicon in [4], also agrees with the theory.

The change in n_{t_0} on the application of a constant field should alter the activation energy of the temperature dependence of τ provided the depletion of the surface is moderate. Thus we can explain the experiments [16] in which the field effect in germanium was investigated at nitrogen temperatures.

The equality (8a) corresponds to the formulas used by Rupprecht [8] to explain his results. The condition for the validity of these formulas, in the form of Eq. (8), may not have been satisfied at the high densities of the surface states occurring in Rupprecht's experiments. Therefore, an interpretation of the experimental data on the temperature dependence of the field-effect relaxation time should include an estimate of the ratio of the excess induced charge and the number of majority carriers in traps.

It follows that the experimental data on the field-effect kinetics in germanium and silicon are in qualitative agreement with the results of calculations.

5. Conclusions

1. Formulas are obtained which describe the field-effect kinetics for the case of large changes of the surface potential when the effect is solely due to majority carriers. It is shown that when the surface states are filled to a considerable degree with majority carriers, the conductivity may depend linearly on time when the space-charge layer is depleted; the relaxation time is then proportional to the induced charge.

2. A new method of determining the surface potential from the field-effect kinetics is proposed. The method is applicable in the case of high surface state densities and the absence of minority carriers at the surface.

3. The results of calculations are in agreement with the experimental data on the field-effect kinetics in germanium and silicon.

The author is deeply grateful to Professor S. G. Kalashnikov for discussing the results of the present work.

LITERATURE CITED

1. A. É. Yunovich, Zhur. Tekh. Fiz. 27, 1707 (1957); Fiz. Tverd. Tela 1, 908 (1959).
2. H. C. Montgomery, Phys. Rev. 106, 441 (1957).
3. V. G. Litovchenko and V. I. Lyashenko, Fiz. Tverd. Tela 3, 61, 73 (1961).
4. V. G. Litovchenko and O. V. Snitko, Fiz. Tverd. Tela 2, 591, 815 (1960); V. G. Litovchenko, Fiz. Tverd. Tela 2, 83 (1959).
5. A. É. Yunovich, Zhur. Tekh. Fiz. 28, 689 (1958); Fiz. Tverd. Tela 1, 1092 (1959).
6. C. G. B. Garrett, Phys. Rev. 107, 478 (1957).
7. G. G. E. Low, Proc. Phys. Soc. (London) B69, 1331 (1956).
8. G. Rupprecht, J. Phys. Chem. Solids 14, 208 (1960); Proc. Intern. Conference on Semiconductor Physics, 1960 (Prague, 1961) p. 28.
9. S. R. Morrison, Phys. Rev. 114, 437 (1959).
10. D. H. Lindley and P. C. Bunbary, J. Phys. Chem. Solids 14, 200 (1960).
11. C. G. B. Garrett and W. H. Brattain, Phys. Rev. 99, 376 (1955).
12. W. L. Brown, Phys. Rev. 100, 599 (1955).
13. A. V. Rzhanov, this volume, p. 70.
14. V. E. Lashkarev, Izvest. Akad. Nauk SSSR, Ser. Fiz. 16, 203 (1952).
15. A. É. Yunovich and V. I. Tikhonov, this volume, p. 97.
16. A. Many, N. B. Grower, Y. Goldstein, and E. Harnik, J. Phys. Chem. Solids 14, 186 (1960).

APPENDIX 1

We shall consider the relaxation of excess charge after switching off the field in the case when there are several surface levels. From Eqs. (2) and (3) it follows that

$$\frac{d\Delta\Gamma_p}{dt} = -\sum_t \alpha_{pt}\left[n_t p_s - p_{s0}\frac{f_{t_0}}{1-f_{t0}}p_t\right]. \tag{16}$$

We shall assume that when the surface is depleted the hole density satisfies the inequalities (6) at every level. It is sufficient to satisfy the condition

$$|\Delta\Gamma_p| \ll \frac{\sum_t \alpha_{pt} n_{t0}}{\sum_t \alpha_{pt}\dfrac{f_{t0}}{1-f_{t0}}} \tag{17}$$

for the relaxation time to be given by the formula

$$\tau_{\text{aver}} = \frac{|\Delta\Gamma_p|}{\left(\sum_t \alpha_{pt}\, n_{t0}\right)\cdot P_{s0}};$$

(17a)

in this case the condition $|\Delta\Gamma_p| \ll \sum_t n_{t0}$ must be satisfied.

When the departure from equilibrium is small, i.e., when the inequalities (10) are satisfied, the condition

$$\left(-\frac{d\Gamma_p}{dy_s}\right)_{v_{s0}} \ll \frac{\sum_t \alpha_{pt} n_{t0}}{\sum_t \alpha_{pt}\, \dfrac{1}{1-f_{t0}}}$$

(18)

is sufficient for the relaxation time to be given by the formula

$$\tau_{\text{small}} = \frac{\left(-\dfrac{d\Gamma_p}{dy_s}\right)_{v_{s0}}}{\left(\sum_t \alpha_{pt} n_{t0}\right)\cdot P_{s0}};$$

(18a)

in this case the condition $\left(-\dfrac{d\Gamma_p}{dy_s}\right)_{v_{s0}} \ll \sum_t n_{t0}(1-f_{t0})$ must be obeyed. If the equalities considered here are applicable, then the value of the derivative $(d\Gamma_p/dy_s)_{y_{s0}}$ — and, consequently, of the surface potential — can be determined from Eq. (15).

APPENDIX 2

To solve numerically the problems of the field-effect kinetics in the cases considered it is convenient to introduce the following dimensionless variables:

$$\overline{\Delta\Gamma}_p = \Delta\Gamma_p / p_0 L_D; \quad \overline{n}_m = n_m / p_0 L_D; \quad \theta = t\cdot\alpha_{pt} p_0.$$

The equation for the function $\overline{\Delta\Gamma}_p(\theta)$ becomes a dependence on three nondimensional parameters describing the semiconductor surface (y_{s0}; $\overline{N}_t = N_t / p_0 L_D$; ε_{t0} is the energy position of the surface level in units of kT, with respect to the Fermi level in the absence of band curvature):

$$\frac{d\overline{\Delta\Gamma}_p}{d\theta} = -\overline{\Delta\Gamma}_p(e^{-\varepsilon_{t0}} + e^{-v_{s0}}) - \overline{N}_t \frac{e^{-v_s} - e^{-v_{s0}}}{1 + e^{\varepsilon_{t0}-v_{s0}}} +$$
$$+ \frac{d\overline{n}_m}{d\theta} + \overline{n}_m(e^{-\varepsilon_{t0}} + e^{-v_s}).$$

(19)

We have used here the fact that $f_{t0} = (1 + e^{\varepsilon_{t0} - y_{s0}})^{-1}$ (the correction to the Fermi distribution of the local levels [5] is neglected). Since y_s depends on $\overline{\Delta\Gamma}_p$ in accordance with Eq. (1), in general we should transform Eq. (19) into the corresponding equation for y_s in order to carry out numerical calculations. The expression for $d\overline{\Delta\Gamma}_p/dy_s$ is obtained by differentiating Eq. (1). When the dependence of y_s on θ is calculated, the quantity $\overline{\Delta\Gamma}_p$ is determined from Eq. (1).

For a completely depleted layer, when y_s, $y_{s0} \gg 1$, we have $y_s = \overline{\Gamma}_p^2$, $y_{s0} = \overline{\Gamma}_{p0}^2$. For an enriched layer, when y_s, $y_{s0} \ll -1$, we have $e^{-y_s} = \overline{\Gamma}_p^2$, $e^{-y_{s0}} = \overline{\Gamma}_{p0}^2$. These simplifications allow us to find $\overline{\Delta\Gamma}_p$ directly from Eq. (19).

If $\overline{n}_m$ is a step function, which was determined in Sec. 3, then for $\theta > 0$ the quantity $\overline{\Delta\Gamma}_p$ is given by an integral similar to that in Eq. (6). In the case of an enriched layer the integral is taken in its explicit form, since the denominator of the integrand is a cubic polynomial in $\Delta\Gamma_p$ (one of its roots is $\Delta\Gamma_p = 0$).

If $\bar{n}_m$ depends on time in accordance with the law $\bar{n}_m = \bar{n}_{m_0} \cos \Omega\theta$, where $\Omega = \omega/\alpha_{pt}P_0$ is a dimensionless angular frequency, then we are interested in the steady-state solution of the problem. This solution can be found by finding the limiting cycle in the phase plane ($\Delta\bar{r}_p$, $d\Delta\bar{r}_p/d\theta$). For comparison with experiment, it is important to know the constant component as well as the first and second harmonics of the Fourier series for this limiting cycle.

KINETICS OF SURFACE PHENOMENA IN SILICON

A. É. Yunovich and V. I. Tikhonov

M. V. Lomonosov State University, Moscow

Relaxation processes with characteristic times of 10^{-1}-10^{-4} sec (at room temperature) were observed in studies of the electrical properties of silicon surfaces using the field-effect method [1-4]. These times are considerably longer than those for the surface recombination of electron—hole pairs but, as a rule, shorter than the times for the establishment of equilibrium between a semiconductor and the atoms and ions adsorbed on its surface.

The nature of these processes is of interest. Rupprecht [5, 6] investigated the conductivity relaxation after the application of an external field pulse. His results could be explained by assuming that the interaction of the surface states only with majority carriers plays the main role in the kinetics of these processes. Morrison [7] regarded the generation of minority carriers at surface states to be important.

The explanation of the field-effect kinetics given in [8-10] applies only to the case of a small departure from equilibrium, while in experiments on the field effect in silicon the departure from electron equilibrium on the surface is, as a rule, large [3, 4].

The purpose of the present work was to study the dependence of the field effect in silicon at room temperature on the frequency and amplitude of the applied field. It was found that the nonlinear variation of the conductivity with increase of the field amplitude at frequencies of 60-2×10^4 cps is related to the dependence of the relaxation time on the band curvature at the surface.

1. Experimental Technique

The frequency dependence of the field effect has already been investigated for germanium [11, 12]. The external fields in these investigations were small so that the changes in the surface potential were much smaller than kT/q. If the surface potential changes are comparable with kT/q, then the nonlinear nature of the variation of the conductivity with the field becomes easily noticeable in the shape of the field-effect signal on the oscillograph screen. In the case of nonlinear periodic variations of the conductivity, the amplitude of the fundamental harmonic of the conductivity change and the phase shift between the fundamental harmonic and the applied external field can be used as quantitative characteristics.

The experimental technique, described in [11, 12], permitted measurements of only the real part of the conductivity change. In our experiment, we used the measuring circuit shown in Fig. 1, which can be used to determine both the real and imaginary components of the field-effect signal. A constant current through the sample is obtained from a bridge circuit with which stray signals, due to displacement currents flowing through the metal–semiconductor capacitor, are balanced out. The alternating voltage resulting from the variation of the sample conductivity is amplified with a differential amplifier. After amplification, the signal is displayed on a cathode-ray oscillograph, and also fed to a phase voltmeter VF-1. The latter measures the real and imaginary parts of this signal with respect to a standard voltage deriving from the same sinusoidal voltage generator which establishes the field on the semiconductor surface. A phase-sensitive detector with a large time constant (≈ 1 sec) at the output of the voltmeter makes the pass band of the instrument sufficiently small for the separation of the fundamental harmonic of the signal.

In other experiments, we measured the relaxation time of the excess conductivity after a field pulse. An exponential scanning voltage with a variable time constant was applied to the horizontal plates of the oscillograph. The relaxation time of the field effect was determined by rectifying the conductivity decay curve on the oscillograph screen, by selecting a suitable scanning time (the tau-meter method). It should be noted that in the case of the pulse method the minimum values of the field at which measurements can still be carried out are several times greater than in the case of the frequency method. The latter, although more complicated to interpret, has advantages in weak fields.

A sample could be illuminated through a semitransparent field electrode either continuously or with chopped light. The semitransparent electrode was a thin layer of tin oxide deposited on a mica plate of 22 μ thickness. A system of light filters made it possible to produce either surface or volume excitation of carriers in the sample. The chopping frequency of the light could be varied between 30 and 1000 cps, and the relative intensity of the illumination could be controlled by measuring the photoconductivity. The photoconductivity signal amplitude was measured with a cathode voltmeter by chopping the light at a low frequency. In the present range of frequencies, this amplitude was independent of the chopping frequency. The relative change of the conductivity $\Delta\sigma_{pc}/\sigma$ due to illumination was 10^{-5}-10^{-2}.

The experiments were carried out in an atmosphere of dry oxygen at room temperature. Results were obtained for four samples, on two of which the measurements were made out in greater detail. P-type samples of silicon had a resistivity of 200-300 $\Omega \cdot$cm and the nonequilibrium carrier lifetime in the interior was 200 μsec. The samples were etched in a mixture of hydrofluoric and nitric acids (CP-8).

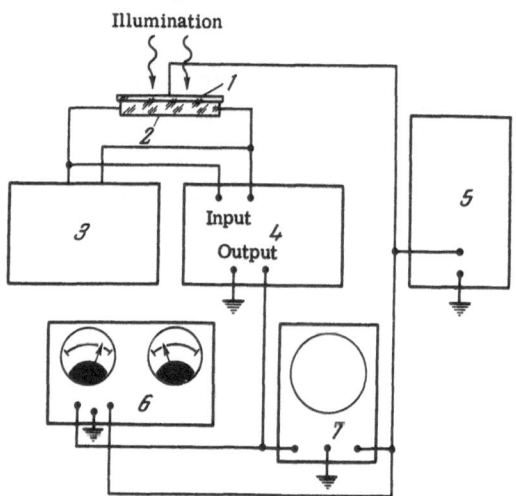

Fig. 1. Circuit for measuring the frequency dependence of the field effect using a phase-sensitive voltmeter VF-1. 1) Semitransparent field electrode; 2) sample; 3) power supply of the sample and compensation system; 4) amplifier with a balanced input; 5) audio oscillator; 6) VF-1; 7) oscillograph.

2. Results of Experiments

1. Typical frequency dependences, in the 60–10^4 cps range, of the real and imaginary components of the field effect in p-type silicon in darkness are shown in Fig. 2. The ordinate gives the effective mobility $\mu_{eff} = \Delta\sigma/\Delta Q$, where $\Delta\sigma$ is the conductivity increment, and ΔQ is the induced charge (per unit area of the semiconductor surface). In general, the concept of effective mobility has meaning only for small signals, i.e., when it represents $d\sigma/dQ$. In the present case, the concept of μ_{eff} is introduced formally as a measure of the conductivity increment in relation to the change of the conductivity due to an external field in the absence of surface states and surface scattering.

The sign of the conductivity and increment $\Delta\sigma$ in all cases represented p-type conductivity at the surface, indicating the absence of an inversion layer at the surface.

The frequency dependence of the imaginary component of the effective mobility $Im[\mu_{eff}(f)]$ had a characteristic maximum; at high and low frequencies $Im[\mu_{eff}]$ was much smaller than at intermediate frequencies. The real component $Re[\mu_{eff}]$ depended weakly on the frequency and was small at low

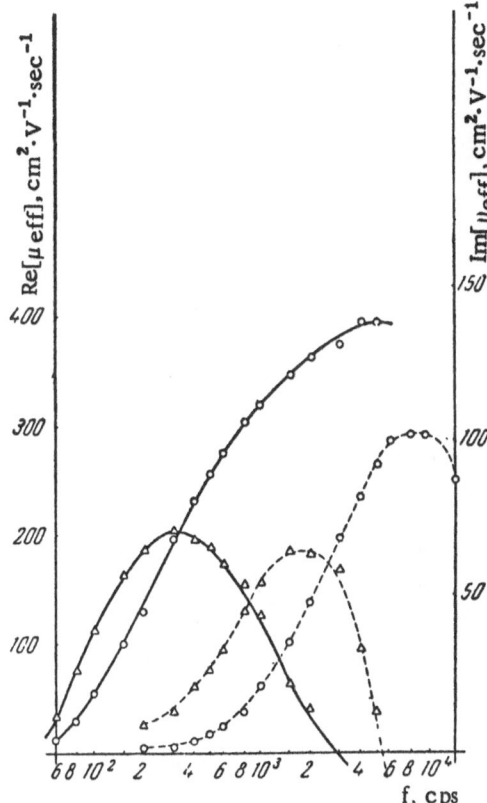

Fig. 2. Typical frequency dependences of the real and imaginary components of the field-effect signal for two different voltages applied to the field plate of the capacitor. —— V_{ac} = 100 V ($\Delta Q = 1.5 \times 10^{-8}$ C/cm^2); ---- V_{ac} = 10 V ($\Delta Q = 1.5 \times 10^{-9}$ C/cm^2); O - Re[μ_{eff}]; Δ - Im[μ_{eff}]. The maximum values of Im[$\mu_{eff}(f)$] correspond to the relaxation times of τ = 560 μsec (V_{ac} = 100 V) and τ = 95 μsec (V_{ac} = 10 V).

frequencies; with increase of the frequency this component rose and tended to saturation.

At first sight, the qualitative form of the curves agrees with the theoretical predictions for the small-signal case [10]. However, quantitative agreement was not obtained: it was impossible to represent the separate dependences Re[$\mu_{eff}(f)$] and Im[$\mu_{eff}(f)$] by the same relaxation time τ. However, to describe qualitatively the processes we can introduce an effective relaxation time as a quantity which is the reciprocal of the angular frequency $\omega = 2\pi f$, at which the value of Im[μ_{eff}] is maximal. The continuous curve in Fig. 2 represents τ = 560 μsec. This value is several times greater than even the volume lifetime of the nonequilibrium carrier pairs in the original material. The lifetime in the samples was considerably shorter since the treatment used produced strong surface recombination.

We should consider the form of the signal representing the conductivity increment. At low frequencies this signal differs strongly from the sinusoid, the change in the conductivity being large during that half-wave of the applied field which corresponds to a depletion of the surface (i.e., minus on the semiconductor). At higher frequencies, at which Im[μ_{eff}] is large, the shape of the signal approaches the sinusoid.

At frequencies of the order of 10 kc the stray signals, due to displacement currents through the metal-semiconductor capacitor, were comparable with the field-effect signal. Whenever the stray effects were considerable, the magnitude of the useful signal was found by averaging two measurements for different directions of the current through the sample.

2. The magnitude of the relaxation time, determined as indicated above, depended on the external field amplitude. An increase of the amplitude shifted the field-effect curves toward lower frequencies, i.e., the effective relaxation time increased. Figure 2 shows the dependence of the real and imaginary components of the effective mobility on the field frequency for two values of the field amplitude differing by a factor of ten.

The variation of the relaxation time with increase of the field amplitude may be due to the variation of the density of either majority or minority carriers at the surface. The hypothesis that the main contribution comes from minority carriers meets with the following objections. First, as just mentioned, the sign of the conductivity increment indicates the absence of an inversion layer. Secondly, the relaxation time of the field effect is much greater than the possible lifetime of the nonequilibrium carrier pairs.

3. Tests on the influence of continuous illumination on the field effect can answer the question: what is the sign of carriers that determine the relaxation processes? Figure 3 shows the frequency dependence of Im[μ_{eff}] for several values of the continuous illumination intensity and of the surface excitation of nonequilibrium carriers. In practice, at these values of ΔV_{pc}, volume excitation did not affect the field effect. The latter circumstance indicates that in the kinetics of surface processes the value of the excess carrier density near the surface is important and not the average value of the nonequilibrium carrier density across the semiconductor thickness, which is important in photoconductivity.

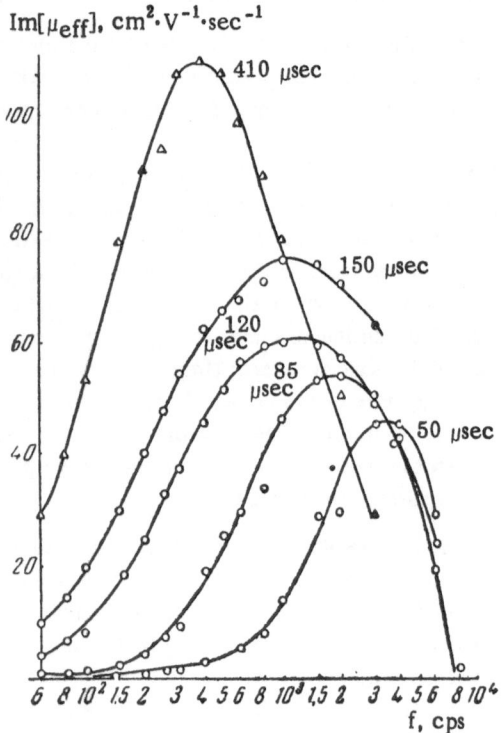

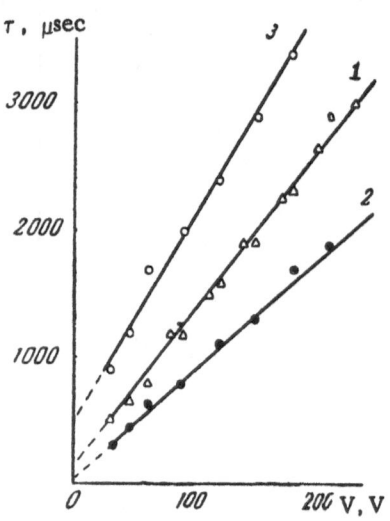

Fig. 3. Influence of continuous illumination on the field-effect kinetics: $\triangle$ - in darkness; $\bigcirc$ - with continuous illumination. Each succeeding period of continuous illumination doubled the photoconductivity voltage $\Delta V_{pc} = I/\Delta\sigma_{pc}$, where I is the current through the sample and $\Delta\sigma_{pc}$ is the change in the conductivity under the influence of continuous illumination. The relaxation times found from the values of the angular frequencies at which maxima of $Im[\mu_{eff}(f)]$ occur were directly proportional to the logarithm of ΔV_{pc}: $\tau \sim -\ln \Delta V_{pc}$.

Fig. 4. Dependence of the relaxation time of the surface conductivity on the amplitude of a negative voltage pulse applied to the field plate: 1) without a constant bias; 2) constant bias of -100 V applied to the field electrode; 3) bias of +100 V.

Stronger continuous illumination reduces the relaxation time and the field effect. The reduction of τ is approximately proportional to the logarithm of the photoconductivity, which is in agreement with the data obtained by Litovchenko [3] by the pulse method. The real component of the effective mobility at low frequencies but under intense continuous illumination changes its sign to n-type conduction, thereby indicating a considerable increase in the number of minority carriers due to continuous illumination.

From this we may conclude that an increase of the field amplitude in darkness does not produce nonequilibrium electrons because the relaxation time of the field effect increases.

4. In experiments using an external pulsed field, the application of a field of the sign corresponding to the surface depletion (positive pulse applied to the plate, i.e., minus on the semiconductor) produced a large change in the conductivity and a slow relaxation to the equilibrium state. If the pulse duration was sufficiently long, the change in the conductivity tended to a value which was too small to be measured ($\mu_{eff} < 5$ cm$^2 \cdot$V$^{-1} \cdot$sec^{-1}). At the end of this pulse, the time of relaxation to the initial surface state was less than 10 μsec. In the region of such short relaxation times after the field pulse, the stray effects were large so that a useful signal could not be separated reliably. The application of a negative field pulse had the same effect as the removal of a positive pulse and, conversely, the removal of a negative pulse was equivalent to the application of a positive one.

The decay of the conductivity increment with time was nearly exponential in the case of the relatively slow relaxation. However, the departure from the exponential law was noticeable, amounting to 10-15%. The relaxation time (determined with a tau-meter) on the application of a positive pulse was somewhat longer than on the removal of a negative one.

If, in addition to a pulse, a constant voltage from a battery was applied to the field electrode, the relaxation time decreased on the enrichment of the surface with majority carriers (minus at the field electrode) and it increased for the opposite polarity. After the application of a constant bias, the relaxation time gradually returned to its initial value after about one hour. The "slow" relaxation processes are usually related to the interaction of the semiconductor with the surrounding medium. In our case (surface of high-resistivity silicon oxidized for a sufficiently long time in an atmosphere of dry oxygen), the slow processes took about one hour. Hence, we may conclude that the relaxation times of the order of hundreds and thousands of microseconds are related to the fast surface state.

An increase of the pulse amplitude increased the relaxation time after the application of a positive, or the removal of a negative, pulse. Figure 4 shows the dependence of the relaxation time of the field effect on the pulse amplitude. The pulse duration was about 8 μsec, the repetition frequency 10 – 40 cps. The upper and lower curves represent measurements carried out with additional constant bias applied to the field electrode: the upper curve corresponds to positive bias, the lower to negative. From the curves, it follows that over a considerable region the relaxation time varies linearly with the field intensity.

5. The nonlinear nature of the photoconductivity variation and the dependence of τ on the field intensity indicate that under the experimental conditions the nonequilibrium change in the surface potential of silicon is greater than or of the order of kT/q. For small signals when the changes in the potential are small compared with kT/q, the relaxation time should be independent of the field amplitude.

A voltage of 100 V applied to the capacitor (in the case shown in Fig. 4) represents a charge of 1.5×10^{-8} C/cm^2. This is larger than the total space charge at the semiconductor surface, $qp_0L_D = 10^{-9}$ C/cm^2 (the hole density in the interior $p_0 = 4 \times 10^{13}$ cm^{-3}; the Debye length is $L_D = 1.4 \times 10^{-4}$ cm). That is why the observed phenomena are nonlinear. Under equilibrium conditions the change in the surface potential under the same external fields may be small because of the strong screening of the field by the surface states.

3. Discussion of Results and Conclusions

1. The results of the experiments show that the proposed method of investigating the kinetics of surface phenomena (measurement of both the real and imaginary components of the field effect) is very effective. Using this method, we can clearly establish the complex nature of the relaxation processes. Since a narrow-band recording circuit is used, measurements can be made at considerably smaller signals than those needed for the pulse method. The latter, however, is simpler and more graphic, and has advantages for large signals.

Quantitative interpretation of the data obtained by both methods in the strongly nonlinear case observed is difficult. The available theories [5-10] do not fit the conditions and results of our experiments.

2. The observed phenomena (the dependence of the relaxation time on the field amplitude, the difference in the effect of different field polarities, the reduction of the relaxation time on continuous illumination) can all be explained by assuming that majority carriers (holes) play the dominant part. In such a case, the theory of Morrison [7], predicting the dependence of the relaxation time on the field amplitude, cannot be used since according to his theory the main cause of the departure from exponential relaxation with a single time constant is the liberation of minority carriers from the surface states.

Rupprecht's theory [5], which takes into account only the majority carriers, is also incapable of explaining the dependence of the relaxation time on the field amplitude; the experimental results show that the possibility of such a dependence should be allowed for. This dependence appears if the experimental conditions are such that the departure from equilibrium is large.

It should be noted that the phenomena observed in germanium at low temperatures [13-15] are very similar to those in silicon at room temperature. Many and Harnik [14] used very large band curvatures, up to 100 kT/q in the carrier depletion direction, but even such curvatures did not produce the liberation of minority carriers in the pulse field effect.

3. The influence of continuous illumination can be explained as follows. Electrons generated by light are obviously captured rapidly by the surface states. The increase of the number of holes near the surface and the simultaneous increase in the electron population of traps accelerate the processes of field-effect relaxation. Under the action of an external field, the electrons captured by the surface states may be liberated by continuous illumination and transferred to the conduction band. This is indicated by the change of the sign of the effective mobility on strong continuous illumination. We note that even very weak continuous illumination, corresponding to a relative change of the conductivity equal to $\Delta\sigma_{pc}/\Delta\sigma \approx 10^{-5}$, alters considerably the value of τ. According to the proposed mechanism, this suggests a considerable change in the population of the surface states, i.e., in the surface charge, even in the case of weak continuous illumination. The results of studies, in which the dependence of the surface recombination velocity on the injection level was investigated [1, 15-18], agree with this conclusion.

4. Yunovich [19] carried out a calculation based on a model which follows from the experiments described above. According to this calculation, there should be a region of linear dependence of the field-effect relaxation time on the induced charge. A high density of the surface states is necessary for the existence of this region. In silicon, etched in a mixture of nitric and hydrofluoric acids and placed in an atmosphere of oxygen, the density of the surface states estimated by various authors [1, 3, 5] is indeed high, being of the order of 10^{12}-10^{13} cm^{-2}. This estimate agrees with the results of our experiments.

The authors express their deep gratitude to Professor S. G. Kalashnikov for his interest in this work and for discussion of the results.

LITERATURE CITED

1. T. M. Buck and F. S. McKim, J. Electrochem. Soc. 105, 709 (1958).
2. M. M. Atalla and M. Tannenbaum, Bell System Tech. J. 38, 749 (1959).
3. V. G. Litovchenko and O. V. Snitko, Fiz. Tverd. Tela 2, 591, 815 (1960).
4. V. G. Litovchenko and O. V. Snitko, Fiz. Tverd. Tela 1, 83 (1959).
5. G. J. Rupprecht, Proc. Intern. Conference on Semiconductor Physics, 1960 (Prague, 1961), p. 282.
6. G. J. Rupprecht, J. Phys. Chem. Solids 14, 208 (1960).
7. S. R. Morrison, Phys. Rev. 114, 437 (1959).
8. V. L. Bonch-Bruevich, Zhur. Tekh. Fiz. 28, 70 (1958).
9. C. G. B. Garrett, Phys. Rev. 107, 478 (1957).
10. A. É. Yunovich, Zhur. Tekh. Fiz. 28, 689 (1958); Fiz. Tverd. Tela 1, 1092 (1959).
11. H. C. Montgomery, Phys. Rev. 106, 441 (1957).
12. A. É. Yunovich, Zhur. Tekh. Fiz. 28, 1707 (1957); Fiz. Tverd. Tela 1, 908 (1959).
13. V. G. Litovchenko and V. I. Lyashenko, Fiz. Tverd. Tela 3, 73 (1961).
14. A. Many, E. Harnik, et al., J. Phys. Chem. Solids 14, 186 (1960).
15. D. N. Lindley and P. C. Bunbary, J. Phys. Chem. Solids 14, 200 (1960).
16. A. H. Benny and F. D. Morton, Proc. Phys. Soc. (London) 72, 1007 (1958).
17. H. M. Bath and M. Cutler, J. Phys. Chem. Solids 5, 171 (1958).
18. Yu. A. Kontsevoi and M. I. Iglitsyn, Fiz. Tverd. Tela 3, 1465 (1961).
19. A. É. Yunovich, this volume, p. 88.

INVESTIGATION OF THE ELECTRICAL
PROPERTIES OF A GERMANIUM SURFACE

V. G. Litovchenko, V. I. Lyashenko, and O. S. Frolov

Institute for Semiconductors, Academy of Sciences, UkrSSR

In the course of investigations of the electrical properties of semiconductor surfaces, it has been established that to determine some very important surface-sensitive effects it is necessary to carry out combined experiments and to develop new methods.

Parallel measurements of the slow relaxation of the conductivity and of the work function facilitate the development of a new method for the study of surfaces. The advantage of this method compared with the usual field-effect method is that one determines experimentally two quantities which characterize the space charge at the surface: the surface conductivity and the surface potential.

In the present contribution, we report the experimental results of a study of the electron structure of a surface by the adsorption of various molecules and by the field effect. The main results were obtained in the study of the slow relaxation of the field effect and the slow relaxation on adsorption.

1. Investigation of the Electron Structure of a Surface by Means of Molecular Adsorption

The investigation of the adsorption and electronic properties of the real surface of germanium were carried out in various ambient gases. We measured the work function φ (the contact potential difference method, using a vibrating electrode) and the conductivity of thin n- and p-type germanium samples having resistivities of 5-55 $\Omega \cdot$cm. Estimates of the change of the work function of a metal reference electrode, $\Delta\varphi_m$, on adsorption in several gaseous environments gave relatively low values (see below).

These experiments yielded the following results.

1. The adsorption of polar and nonpolar molecules on the real surface of germanium strongly affects its electron work function. The value of the change of φ depends on the gas pressure p. At high p (but less than the saturation vapor pressure p')$\Delta\varphi(p)$ has one or more plateaus which obviously represent the completion of a monolayer (Figs. 1 and 2).

Using the value of $\Delta\varphi$ corresponding to a plateau, we can calculate the charge density produced by one monolayer: $n_s^0 \approx 5 \times 10^{11}$ cm^{-2}, which is close in its order of magnitude to the estimate of n_s^0 from $\Delta\sigma$. Thus we can estimate the charge per one adsorbed atom: $e_s = (n_s^0 / n_{ads}) \times e \approx 10^{-3}$-$10^{-4}$e. Such a small value of e_s has been reported in all the papers on germanium and on other semiconductors [1-5]. There are several possible interpretations of this observation: weak electron binding, various quantum-mechanical systems, nonionic binding, partial saturation with electrons.

The value of $\Delta\varphi$ differs greatly for different adsorbates; it depends on the state of the surface and even on the resistivity of the sample [4]. For example, in the case of ground samples and samples oxidized by heating in oxygen φ_0 was larger and $\Delta\varphi$ several times smaller than for etched samples (Fig. 1). On the other hand, several workers have shown [6, 7] that the specific adsorption of Ge, GeO_2 and even Si, SiO_2 differ only little (for $\theta < 1$). Therefore, the differences in $\Delta\varphi$ are mainly due to the influence on $\Delta\varphi$ of the initial work function φ_0 and the band curvature at the surface Y_s^0 (which differ very considerably for the surface states mentioned above), and not due to differences of n_{ads}. A qualitative analysis of $\Delta\varphi$ can be carried out on the basis of the theory of the space charge near the surface [4, 8]. Asymptotic formulas were obtained for $\Delta\varphi(p)$ (Figs. 1 and 2).

2. It was found that the nature of the adsorbed molecules determines the sign of the change of φ and σ on adsorption and influences the magnitude of $\Delta\varphi$; the sign of the change is not governed by the sign of the end of the dipole which is adsorbed on the surface. Molecules with positive polarity of the dipole (H_2O, ethyl alcohol), negative polarity of the dipole (acetone and CO), nonpolar molecules (O_2) and various mixtures of the constituents of air (dry air, humid air, etc.) were used. On the adsorption of both polar and nonpolar molecules, the value of φ decreased. The exception was O_2, and in some cases CO (low pressures $p \ll 0.1$ mm Hg and fresh surface outgassed by heating in vacuum) which increased φ. Thus O_2 forms a negative charge on the surface, acting as an acceptor. The adsorption of CO affects φ in a complex manner (the complex nature of the influence of adsorption on φ and σ was reported for several cases, for example, for the adsorption of O_2 on atomically clean surface of germanium [9, 10]). The remaining molecules all produced positive charge, i.e., they acted as donors. The sign of $\Delta\sigma$ depended also on the type of conduction of the sample and it was in agreement with the sign expected from $\Delta\varphi$. These observations for the real surface of germanium were found to be similar to our earlier results for such semiconductors as Cu_2O, CuO, ZnO, and others. These observations are easily interpreted if we assume that the change of φ is related to the change of the space charge at the surface, i.e., the band curvature at the surface Y_s.

3. Experiments were carried out both at room temperature and at higher temperatures T. It was found that with the increase of T the value of $\Delta\varphi$ decreased considerably but the sign of $\Delta\varphi$ was not affected. The exception was O_2, for which $\Delta\varphi$ increased with increasing temperature and irreversible changes of the surface properties (φ_0, σ_0, $\Delta\varphi$, and others) were observed. The reduction of $\Delta\varphi$ was ascribed to the reduction in the number of adsorbed molecules. A study of $\varphi(T)$ in vacuum in the range of T of interest to us (~ 300-$430°$K) showed that the change of the Fermi level in the interior, μ_F, and on the surface, μ_{Fs}, do not affect $\Delta\varphi$; usually φ_0 and μ_F (i.e., also Y_s) are approximately constant in this range of temperatures [4].

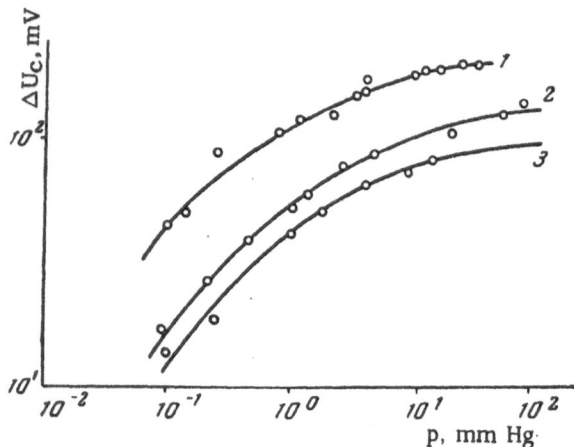 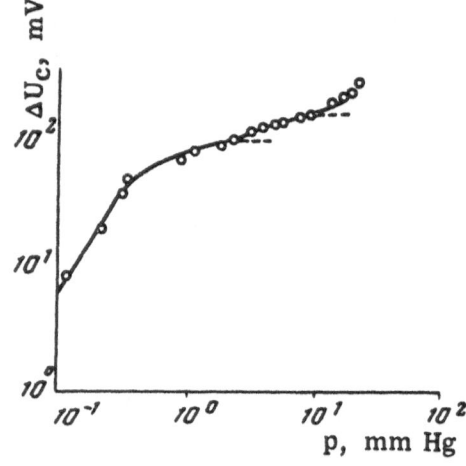

Fig. 1. Variation of the contact potential difference with the pressure of adsorbed CO gas on n-type Ge. 1) Surface cleaned by vacuum heating; 2) surface not cleaned by vacuum heating; 3) surface oxidized by heating in O_2; continuous curves 1 and 2 were calculated from $p = c_1 \times \exp(\Delta U_c / 2m)$; continuous curve 3 represents the formula $p = c_2 \exp[\exp(\Delta U_c / 2)]$.

Fig. 2. Variation of the contact potential difference with the pressure of adsorbed ethyl alcohol vapor on p-type Ge (the continuous curve was calculated theoretically for $p_{\theta=1} = 2.5$ mm Hg, $p_{\theta=2} = 10$ mm Hg).

The effective activation energy for $\Delta\varphi(T)$ at $p = 4$ mm Hg was 0.1 eV $\approx$ 2.2 kcal/mole for ethyl alcohol, 0.14 eV $\approx$ 3.4 kcal/mole for CO, 0.08 eV $\approx$ 2 kcal/mole for acetone, which was indicative of physical adsorption. For oxygen, the activation energy was considerably greater (over 6 kcal/mole at $p = 0.1$ mm Hg) and $\Delta\varphi$ rose with increasing T, which was characteristic of chemisorption. It is interesting to note that these results agree with the direct experiment on adsorption characteristics [6]. We are of the opinion that it would be useful to obtain even qualitative data from the adsorption characteristics in these cases when the direct methods are difficult to use (samples with a small working area, high p, etc.).

4. The equilibrium values of φ and σ after adsorption on Ge (irrespective of whether polar or nonpolar molecules were adsorbed) were established very slowly ($\sim$10-10^3 sec). On the other hand, the experiments on samples of CuO, Cu$_2$O, ZnO, CdO, etc. [1, 2, 4] showed that usually the equilibrium values of φ and σ were established very rapidly (in less than 1 sec) on adsorption. This rapid process is also expected theoretically for physical adsorption [11]. Thus, it seems that the slow variation of φ on germanium is due to the establishment of a new electron equilibrium between the surface and the interior after the completion of adsorption.

The most characteristic features of the kinetics are as follows.

(i) As a rule there was no marked change of $\Delta\varphi(t)$ during the first few seconds (Figs. 3 and 4). For small values of $\Delta\varphi < 1kT/e = 25$ mV the dependence $\Delta\varphi(t)$ was always simple exponential with a time constant τ. The departure from the exponential at higher values of $\Delta\varphi$ (higher p) was related to the departure from the linear conditions in the "small-signal" theory [12]. These observations indicate that φ does not change immediately after adsorption, i.e., the adsorbed molecules are locked at the surface in a practically neutral state and only gradually they acquire a charge which, in its turn, alters Y_s. The same observations also indicate that the change in the electron equilibrium occurs not due to the change in charge but due to the change in the density of surface states on the adsorption of molecules. Moreover, the absence of a marked jump of σ and $\Delta\varphi$ as well as the small value of e_s tend to disprove the ionic binding of molecules with the Ge surface and favor ionic-covalent type of binding with a relatively low proportion of ionicity.

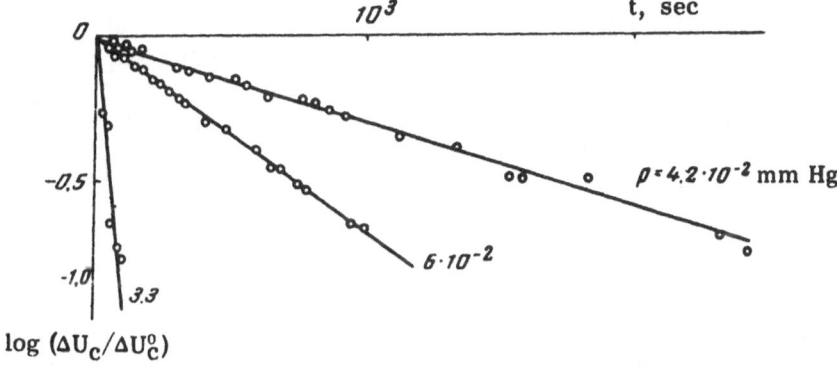

Fig. 3. Dependence of the change in the contact potential difference on time (on semilogarithmic scale) during the adsorption of CO on n-type Ge.

It is not possible to describe $\Delta\varphi(t)$ in terms of functions characteristic of the adsorption processes:

$$\Delta\varphi = (1 + \alpha t)^{-n}, \quad \alpha t^{-n}, \quad A + B \ln(t + t^*),$$

The exception is O_2 and humid air, particularly when acting on an outgassed surface at high T, where $\Delta\varphi = A - B \ln(t + t^*)$ [4].

(ii) The magnitude of τ depended strongly on the thickness of the surface oxide layer d_{ox} [3, 4], indicating electron transport between the surface and the interior through the oxide layer by the tunnel effect. The dependence $\tau(d_{ox})$, obtained using our technique for the determination of τ, is very important for a quantitative comparison with the theory of electron transport between the outer (surface) levels and the interior.

(iii) The value of τ depended strongly on p, decreasing with increase of p: $\tau = \alpha p^m$, where $m \approx 0.5$ (Fig. 5) and was practically independent of the nature of the molecules, of d_{ox}, T, and the resistivity ρ.

(iv) The value of τ rose strongly with increase of temperature:

$$\tau = \tau_0 e^{-E_\tau/kT},$$

where E_τ was different for different types of molecule but the same for the same type of molecule adsorbed on different samples (cf. the table).

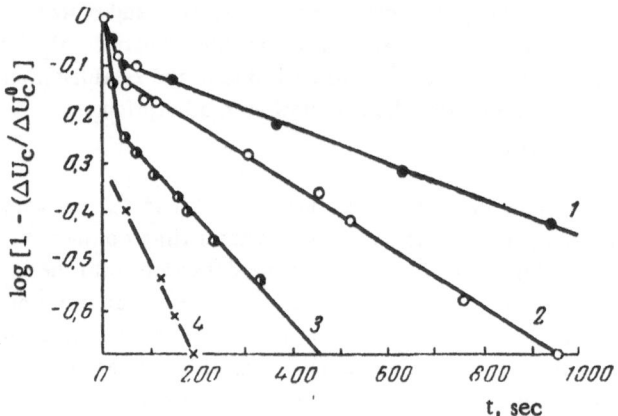

Fig. 4. Dependence of $\log[1 - (\Delta U_c/\Delta U_c^0)]$ on time during the adsorption of ethyl alcohol vapor on n-type Ge at various pressures: 1) $p = 10^{-2}$ mm Hg; 2) $p = 10^{-1}$ mm Hg; 3) $p = 4 \times 10^{-1}$ mm Hg; 4) $p = 1$ mm Hg.

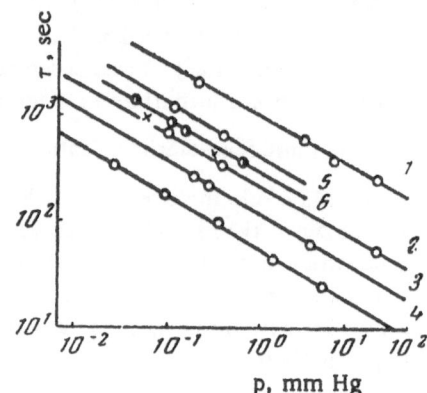

Fig. 5. Dependence of the intrinsic slow-relaxation time on pressure. 1)-4) Adsorption of acetone vapor: 1) $T = 20°C$; 2) $T = 83°C$; 3) $T = 118°C$; 4) $T = 156°C$. 5) and 6) Adsorption of other gases: 5) O_2; 6) CO. The sign $\times$ denotes molecules of dry air at $T = 20°C$.

Apart from the foregoing observations, the activation-adsorption nature of the slow kinetics is also contradicted by a) the reduction of $\Delta\varphi$ with increase of T; b) the strong increase of τ with increase of d_{ox} while the specific adsorption of the surface is not greatly affected.

Thus, the slow changes of $\Delta\varphi$ on adsorption are governed by the establishment of a new electron equilibrium between the outer surface levels (known as "slow" levels) and the interior of the semiconductor. As previously mentioned, the departure from equilibrium is related to the change in the density of these levels, i.e., new "impurity" levels should be formed on adsorption. The kinetic data confirmed this. The activation energy of the kinetic process E_τ includes, in principle, the height of the barrier on the surface and the energy depth of the level, being governed by the larger of these two quantities. The fact that E_τ is the same for one type of molecule on surfaces which are initially different (cf. the data in the table for E for various samples with different Y_s^0, φ_0; cf. the table) indicates the lack of influence of the barrier on the value of E_τ. However, E_τ differed strongly for different types of adsorbed molecule (experiments on the same sample using different types of molecule were carried out under the same initial conditions: φ_0, $Y_s^0 \approx$ const.). This means that on adsorption there is a change only in the surface level (one level, as indicated by the simple exponential nature of the kinetics under these conditions) which is related to the particular type of adsorbed molecules. Otherwise, the difference in the initial Y_s (in experiments on different samples) would have filled different levels each time, giving different E_τ.

Such "slow" surface levels should be produced on adsorption because the initial system of levels is found to be the same for several types of molecule on one sample, and all the initial levels are equivalent (since the surface charge is not altered immediately after adsorption, as pointed out above). Therefore, there is no preference for the change of population of any particular level after adsorption, but the values of E_τ are different depending on the type of molecule.

TABLE 1

Sample No.	Conduction type	ρ, ohm·cm	Adsorbed molecules	E_T, eV
4	n	5	Ethyl alcohol	0,32
5	n	32	Ethyl alcohol	0,31
2	p	38	Ethyl alcohol	0,30
4	n	5	Acetone	0,23
6	p	55	Acetone	0,22
5	n	32	Acetone	0,22
4	n	5	CO	0,09

(v) Thus the adsorption of various types of molecule on the surface of germanium produces "slow" surface levels, which are "impurity" in nature. Molecules of ethyl alcohol produce deep donor levels (E = 0.32 eV), while acetone and carbon monoxide produce shallow donor levels (0.22 eV and 0.1 eV, respectively). Molecules of O_2 produce acceptor levels. The change of φ_{sc} (where "sc" denotes semiconductor) is related to the change of the space charge at the surface, i.e., the surface potential, the geometrical dimensions of which ($\approx \varkappa^{-1}$) are usually much greater than the length of the molecular dipole l_d. In this case the dipole can be regarded as a point charge and its effect on φ_{sc} as the influence of a local center on the charge in the surface region of the semiconductor. If l_d becomes comparable with $\varkappa^{-1}$ (for example, in the case of adsorption of organic dye molecules on CuO, which is a semiconductor with a small $\varkappa^{-1} \approx 10^{-6}$ cm) then, as expected, the sign of the dipole determines the sign of $\Delta\varphi_{sc}$ [14]. Moreover, the influence of the adsorption of molecules on the value of φ is different for a real and for an atomically clean surface because of the large separation of adsorbed molecules from the crystal by an oxide layer on a real surface and because on a real surface molecules of many substances are physically adsorbed. We may assume that, in general, all the "slow" levels on a real surface of Ge are due to adsorbed molecules. It follows from the above discussion that on a real surface there can be a great variety of these levels and this provides us with the possibility of controlling the electrical properties of the surface. Theoretically, the existence of "impurity" surface levels of this type was predicted in the series of papers by Vol'kenshtein et al. [15-17]. The experimental confirmations were also obtained by other methods [12, 13, 17, 18].

(vi) The strong dependence of τ on d_{ox} and on the pressure of the ambient gas, as well as the exponential nature of the $\Delta\varphi$ kinetics, indicate that the "slow" surface levels are on the outer surface and there are few states in the oxide layer.

2. Investigation of the Germanium Surface Using the Slow Relaxation of the Conductivity and of the Work Function in the Field Effect

The experimental technique was as follows.

A constant external electric field of $(2-7) \times 10^5$ V/cm was applied for 1-2 min to the surface of a germanium sample ($15 \times 10 \times 0.3$ mm) through mica. After the application of the field, the "field" electrode was quickly removed with a special device and a platinum vibrating plate electrode was brought to the surface to measure the change in the contact potential difference ΔU_C and in the sample conductivity $\Delta\sigma$ which gradually decreased to zero. Although during the measurement of one of the quantities ($\Delta\sigma$ or ΔU_C), the electrical circuit for the measurement of the other quantity was disconnected, the slow variation of $\Delta\sigma$ and ΔU_C after removal of the external field V_e made it possible to measure both time dependences, $\Delta\sigma(t)$ and $\Delta U_C(t)$, for the same cycle. The construction of the apparatus facilitated the measurement of the contact potential difference and the conductivity 5-10 sec after the removal of the external field. The following ambient atmospheres (apart from the normal atmosphere) could be obtained: a stream of dry argon, a stream of moist argon (75% humidity and saturated vapor), saturated acetone vapor in a stream of argon.

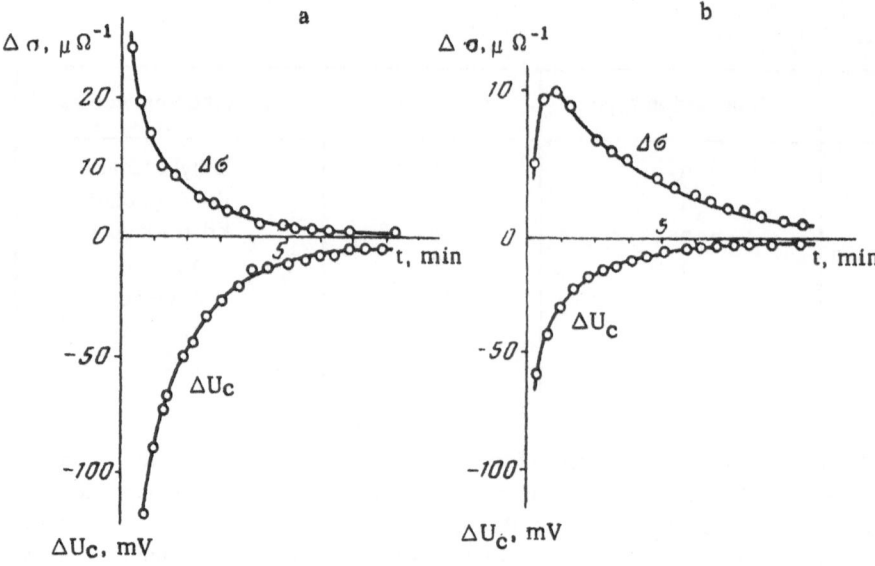

Fig. 6. Dependence of the conductivity on time after switching off an external
field of a) +1500 V and b) -1500 V.

The test sample was cut from a single-crystal ingot of n-type germanium having a resistivity of $\rho \approx 50$ $\Omega \cdot cm$. The sample surface was oriented in the (111) plane and the current direction approximately coincided with the (110) axis. Two current contacts and two point-probe ohmic contacts of tin were attached to the back surface. After soldering the electrodes, the sample was etched in boiling Perhydrol for about 2 min, washed in water and then in ethyl alcohol. Before the measurements, the sample was stored in air for about 1.5 months.

Figure 6 shows typical dependences of the changes $\Delta\sigma$ and ΔU_c on time after the application of a constant transverse electric field V_e of positive (Fig. 6a) and negative (Fig. 6b) polarity at the semiconductor. Similar dependences were obtained for various values of V_e, various gaseous environments, etc. The ΔU_c curves showed usually a monotonic decrease. As expected, the induced positive charge on the semiconductor surface raised the value of the work function φ, and when the field was removed this charge reduced φ. The re-establishment of the initial value of φ_0 usually occurred in several minutes. The time for the re-establishment of φ_0 decreased with increase of the relative humidity of the ambient gases; it increased with the duration of storage in air (i.e., with increase of the thickness of the surface oxide layer), and in acetone vapor.

Figure 7 gives the experimental data, plotted as $\Delta U_c(\log t)$. Dependences of this type were also obtained by Pratt and Kolm [12] and they found that the experimental results obeyed the law

$$\Delta U_c = A - B \ln t. \tag{1}$$

The above relationship can also be obtained theoretically under certain assumptions [12]. It follows from Fig. 7 that in a certain time interval our results also agree with Eq. (1). However, for very long periods ("small signals") and for short periods after switching off V_e there was a departure from the relationship of Eq. (1). The experimental results for longer t, when $\Delta\varphi$ becomes small (≤ 20-25 mV ≈ 1 kT/e), always obeyed the simple exponential law (Fig. 7). The time constant τ of the exponential was independent of the amplitude and polarity of the external field. The simple exponential nature of the time dependence for small $\Delta\varphi$ (both for the field effect and the adsorption) supports the homogeneous model of the surface. The value of τ rose considerably (from 100 to 300 sec) with increase of the oxide layer thickness.

In the case of short times, a stronger dependence of ΔU_c on time was observed than that which follows from Eq. (1). It was found that the theory also predicts, instead of Eq. (1), a stronger dependence for "large signals" (when we cannot limit ourselves to linear terms in the expansion) and an exponential dependence for "small signals."

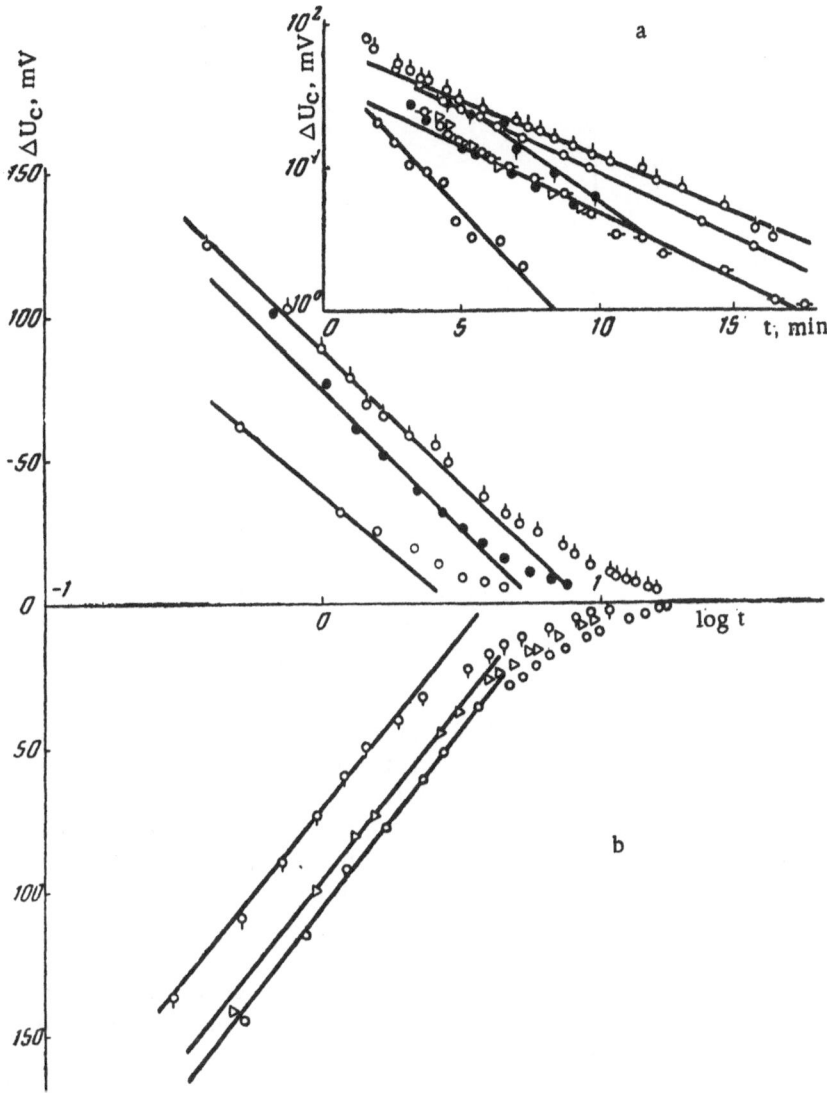

Fig. 7. Dependence of ΔU_C on the duration of storage in air after switching off a transverse field: a) dependence of ΔU_C on time; b) dependence of ΔU_C on the logarithm of time.

Some important characteristics of the surface can be obtained by a detailed analysis of the dependence $\Delta\varphi(t)$ [12, 13].

The relaxation curves $\Delta\sigma(t)$ usually show a monotonic decrease, but in some cases (after switching off a positive V_e or switching on a negative one) they were in the form of curves with a minimum (Fig. 6) due to the passing of the quasi-surface conductivity (i.e., the conductivity of the space-charge layer) through a minimum. In the case of very humid media the value of σ is also affected (in addition to the quasi-surface conductivity) by the conductivity of a condensed liquid film [19-21], which complicates greatly the interpretation of the results. The analysis of such results is not given here.

After eliminating time from the curves $\Delta\sigma(t)$ and $\Delta U_C(t)$, we obtain the dependences $\Delta\sigma(\Delta U_C)$ (Fig. 8). In most cases, the change $\Delta\sigma$ can be related to the change of the quasi-surface conductivity (with the exception of environments containing large amounts of moisture). The quantity ΔU_C, equal to the change in the

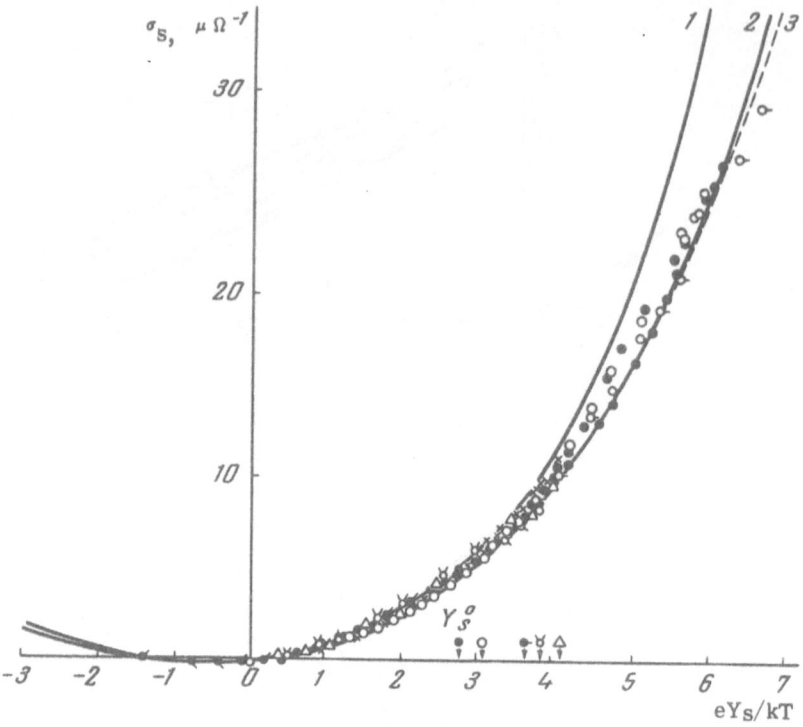

Fig. 8. Dependence of the quasi-surface conductivity on the change in the band curvature at the surface (continuous curves 1, 2, and 3) and in the contact potential difference (experimental points). 1) Theoretical dependence based on the Brown – Garrett – Brattain theory [8]; 2) theoretical dependence with an allowance for surface scattering and using $m^* = 0.117\ m_e$; 3) as 2) but for $m^* = 0.25\ m_e$. The experimental results were obtained in air on different days.

work function of the semiconductor, consists in principle of the following components: ΔY_s, which is the change in the band curvature at the surface, and ΔV_{ox}, which is the change in the potential drop across the oxide layer. Moreover, there may be a potential drop in the dipoles of the adsorbed molecules but this does not affect greatly the change ΔY_s (i.e., also that of σ_s). Here the important contribution is not the potential drop in the dipoles but the absence of a possible slow (with a time constant $\tau_1 \approx \tau$) reorientation or slow desorption after the external electric field is removed. We should note that when $\tau_1 \gg \tau$ there should be a residual effect on $\Delta\sigma$ depending on V_e, but this was not observed experimentally.

When the dependence $\Delta\sigma_s(Y_s)$ was obtained by the field-effect relaxation method, the influence of the potential drop in the dipoles of the adsorbed molecules was checked by recording this dependence in various environments: one with positive dipoles (humid air), another with negative dipoles (acetone), and a third with no dipoles (dry argon). The $\Delta\sigma(\Delta U_c)$ dependences obtained in these environments were practically identical. Thus the influence of the potential drop in the dipoles could be neglected. There are several observations, in particular those described in Sec. 1, which can be used to prove that in general the potential drop in the adsorbed molecular dipoles does not affect greatly the value of φ_{sc}. Among these, are the absence of a correlation between the sign of the molecular dipole end which is in contact with the surface and the sign in the change of the work function of the semiconductor – in contrast to metals, for which there is always such a correlation. True, the absence of such a correlation does not, strictly speaking, prove the absence of the effect of the dipole field on the value of φ of a semiconductor. However, this absence indicates unambiguously that the dipole field does not play the dominant role (with the exception of molecules with very long dipoles) and has no effect whatever in the case of nonpolar molecules (O_2, H_2, N_2, and probably CO, O_3, etc.).

In order to decide in which cases the change in the potential drop across the oxide, ΔV_{ox}, is unimportant compared with ΔY_s in the experimentally determined quantity $\Delta U_c = \Delta \varphi_{sc} = \Delta Y_s + \Delta V_{ox}$, we carried out the appropriate calculations and experiments. The calculations showed that for intrinsic germanium $V_{ox} \approx 0.4$ kT/e at 300°K and $Y_s = 7$ kT/e; V_{ox} was considerably smaller when Y_s was smaller. Thus, in our case, the relative contribution of V_{ox} to ΔU_c was small (less than 5%) throughout practically the whole range of the Y_s values investigated.

If the V_{ox} contribution to ΔU_c had been important, then the form of the $\Delta \sigma(\Delta U_c)$ curves should change considerably with the change in the oxide layer thickness d_{ox}; conversely if $\Delta V_{ox} \ll \Delta Y_s$, then the dependence $\Delta \sigma(\Delta U_c)$ should remain unaltered. Two methods of increasing the oxide thickness were employed: 1) the natural growth of the oxide on prolonged storage of the sample in air; and 2) oxidation by prolonged heating in air (30 hours at 110°C). The dependences $\Delta \sigma(\Delta U_c)$ taken on different days (which corresponded to different d_{ox}) were identical. On the other hand, the curves obtained after the forced growth of the oxide (by heating) were identical with the curves before heating only up to $Y_s \approx 5.5$ kT/e and after that the dependences $\Delta \sigma(\Delta U_c)$ became flatter, indicating a considerable contribution from ΔV_{ox} to ΔU_c. Repeated heating gave an even flatter curve. Thus, after heating, the effect of the oxide on $\Delta \sigma(\Delta U_c)$ for large Y_s became noticeable. The dependences $V_{ox}(Y_s)$ were determined. They were governed by several surface parameters: d_{ox}, ε_{ox}, and the density and energy positions of the surface levels. Calculations showed that the increase of V_{ox} on heating was related both to the increase of d_{ox} and to the increase of the density of "fast" surface levels [4].

The experimental results were compared with the theory of surface conductivity and surface scattering of carriers. In Fig. 8, curve 1 represents the theoretical dependence $\sigma_s(Y_s)$ obtained from the theory of the quasi-surface conductivity $|\sigma_s|$ propounded by Brown, Garrett, and Brattain [8]. It is seen that at low band curvatures Y_s the experimental results are in good agreement with the theory, while at high Y_s there is a systematic departure from the theoretical curve, the experimental points lying below that curve. This is to be expected in the case when the carriers suffer additional scattering at the surface [22]. Allowance for the surface scattering gave the theoretical curve 3 (for the usually accepted value of the effective mass $m^*/m_e = 0.25$) or curve 2 [for $m^* = 3(m_1^{-1} + m_2^{-1} + m_3^{-1})^{-1} = 0.117 m_e$] [22]. As shown in Fig. 8, the experimental results are closer to the theoretical curve which allows for scattering than to the curve which makes no such allowance. However, the results obtained lie on the average above the curve plotted allowing for the Schrieffer correction. From this alone, we cannot conclude that Schrieffer's basic assumption of completely diffuse scattering is not satisfied, but we do have partial diffuse scattering. It is necessary to allow for the anisotropy of surface scattering when $Y_s \neq 0$ [22]. The question is now how important is this anisotropy in our case. Zemel and Petritz [23] calculated the anisotropic surface mobility making certain assumptions, well justified for germanium. Using these assumptions we plotted $\mu_s(Y_s)$ for the principal crystallographic directions (Fig. 9). It is seen that the discrepancy in μ_s is particularly large at $Y_s \geq 4\text{-}7$ kT/e.

The same figure shows the experimental values of μ_s, calculated for the corresponding values of Y_s from the relationship

$$\sigma_s = e\mu_{ps}P_s + e\mu_{ns}N_s,$$

where σ_s, P_s (the excess hole density at the surface) and N_s (the excess electron density at the surface) are known for each value of Y_s. The experimental points are, as a rule, near the theoretical curves 2 (the isotropic case) and particularly 3, which corresponds to the current direction along (110). Thus these results are a direct proof of the importance of surface scattering and they support the diffuse nature of carrier scattering at the surface of germanium combined with a small amount of specular carrier reflection. To determine quantitatively the degree of diffuseness of scattering, experiments should be carried out on a sample oriented accurately along various crystallographic directions.

The proposed method makes it possible to determine the magnitude of the initial band curvature at the surface even without determining the minimum of σ_s. From the experimentally determined dependence $\Delta \sigma_s(Y_s)$, we can find the value of Y_s for each σ_s, including the initial conditions. Then even if there is no minimum of σ_s, the value of Y_s^0 can be obtained by superposition of the experimental curve $\Delta \sigma(\Delta U_c)$ and one obtained earlier, or the theoretical curve. The values of Y_s^0 obtained in this way and by the field-effect method were in good agreement (usually the discrepancy was ≤ 0.2 kT/e).

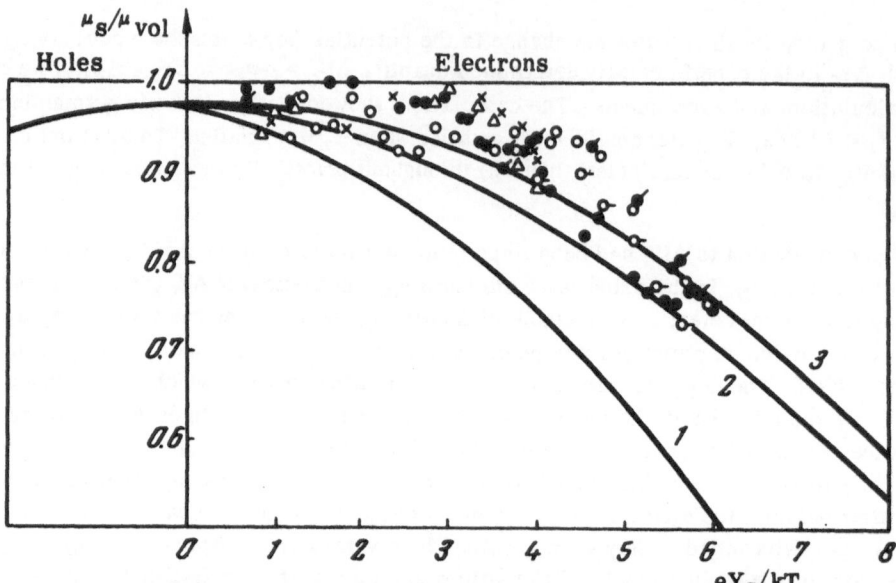

Fig. 9. Dependence of the effective carrier mobility in the surface-charge region on Y_s. Points represent the experimental results; curves were calculated theoretically: 1) for the (111) direction; 2) for the isotropic case; 3) for the directions (100), (110). The results were obtained on different days during storage in air.

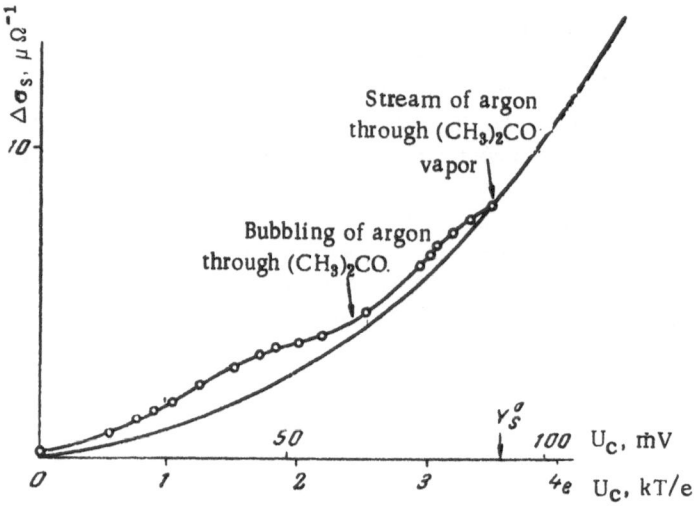

Fig. 10. Dependence of the change in the conductivity on the surface potential: — experimental data from the field effect in a stream of argon; o-o-o from the adsorption of acetone $(CH_3)_2CO$.

The method proposed can be used also to determine the influence of the adsorption and desorption of molecules on the work function of the metallic reference electrode. As found above, the form of the dependence $\Delta\sigma(\Delta U_c)$, obtained by the method of slow relaxation of the field effect, is independent of the surrounding gas. The $\Delta\sigma(\Delta U_c)$ curves can also be obtained by a different method: changing the ambient medium (going over from air to dry acetone vapor in a stream of argon), as shown in Fig. 10. The discrepancy between the curves obtained by these two methods is due either to the change of φ_{sc} by the double electric layer of acetone molecular dipoles, or the change of the work function of the metal on adsorption, since both these effects are

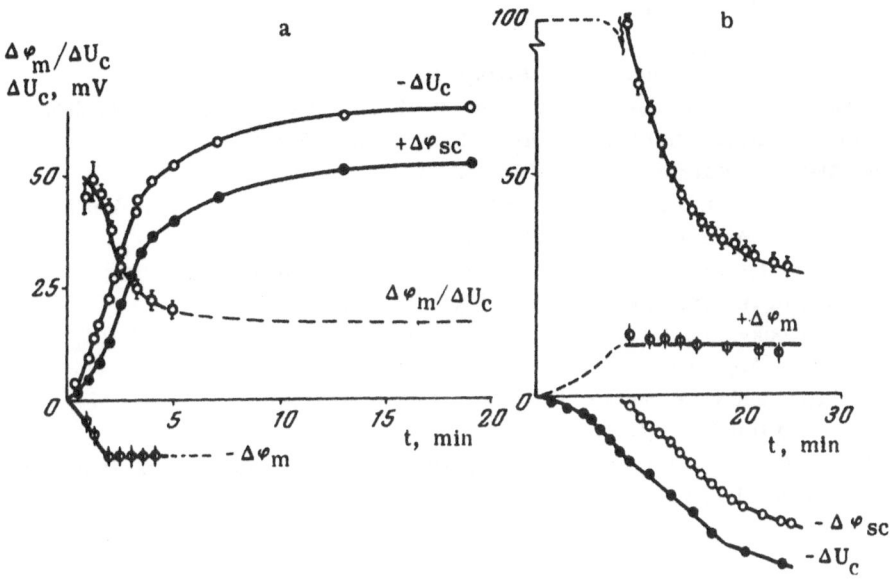

Fig. 11. Variation of U_c, and of the work function of a semiconductor φ_{sc} and a metal φ_m with time during adsorption (a) and desorption (b) of acetone molecules in a stream of argon.

absent in the field-effect measurements of $\Delta\sigma$ and ΔU_c. We may assume (see above) that on the adsorption of polar molecules on the surface of germanium the work function of the latter changes solely due to the change of the "macrodipole," i.e., the change in the quasi-surface layer in the semiconductor. In that case, we can neglect the dipoles of the molecules themselves. An additional proof of this is provided by the identity of the sign of the change of the work function φ of the metal and the predicted change (see below). Thus the discrepancy between the curves in Fig. 10 is due to the change of the work function of the reference electrode on adsorption of acetone molecules. From the difference between these two curves along the abscissa axis, we can determine the change in the work function of the metal φ_m at various ΔU_c, i.e., for different moments during adsorption (Fig. 11). The sign of $\Delta\varphi_m$ agrees with the observation that acetone molecules are adsorbed as a positive charge, while water molecules are adsorbed as a negative charge, etc. From Fig. 11 it follows that the value of φ_{sc} is established very slowly while that of the metal is established relatively quickly. As a result of this, in spite of the small value of φ_m, the total effect ΔU_c immediately after adsorption may be determined to a great extent by the change in φ_m. This is illustrated in Fig. 11 by the dependences $\Delta\varphi_m/\Delta U_c(t)$. When the times for establishment of $\Delta\varphi_{sc}$ and $\Delta\varphi_m$ differ considerably we can, in principle, determine these quantities separately by measuring the kinetics of ΔU_c.

LITERATURE CITED

1. V. E. Lashkarev and V. I. Layshenko, in the book: Collective Volume Dedicated to Academician A. F. Ioffe on the Occasion of his 70th Birthday (Izd. AN SSSR, 1950).
2. V. I. Lyashenko and I. I. Stepko, Zhur. Fiz. Khim. 29, 401 (1955).
3. V. I. Lyashenko and V. G. Litovchenko, Zhur. Tekh. Fiz. 28, 454 (1958).
4. V. G. Litovchenko, Dissertation for Candidate's Degree (Institute for Semicondcutors, AN UkrSSr, Kiev, 1960).
5. G. Dorda, Czech. J. Phys. B10, 820 (1960).
6. J. T. Law, J. Phys. Chem. 59, 67 (1955).
7. G. A. Romanova and I. I. Stepko, this volume, p. 56.
8. W. Brown, Phys. Rev. 100, 590 (1955); C. G. B. Garrett and W. H. Brattain, Phys. Rev. 99, 376 (1955).

9. P. Handler and W. Portnoy, Phys. Rev. 116, 516 (1959).

10. R. Forman, Phys. Rev. 117, 698 (1960).

11. J. de Boer, The Dynamical Character of Adsorption (Oxford, 1953), p. 13.

12. G. W. Pratt and H. H. Kolm, Semiconductor Surface Physics (New York, 1957), p. 297.

13. S. R. Morrison, Semiconductor Surface Physics (New York, 1957), p. 169.

14. V. I. Lyashenko, Doctoral Dissertation (Kiev, 1955).

15. F. F. Vol'kenshtein, Zhur. Fiz. Khim. 21, 1317 (1947); 26, 1462 (1952); F. F. Vol'kenshtein and V. B. Sandomirskii, Problemy Kinetiki i Kataliza 8, 189 (1955).

16. V. L. Bonch-Bruevich, Zhur. Fiz. Khim. 26, 1462 (1952); 27, 662 (1953).

17. H. J. Engell and K. Hauffe, Z. Elektrochem. 57, 562 (1953).

18. M. Lasser, J. Wysocky, and B. Bernstein, Phys. Rev. 105, 491 (1957).

19. J. T. Law et al., J. Appl. Phys. 27, 1265 (1955).

20. V. G. Litovchenko and O. V. Snitko, Fiz. Tverd. Tela 2, 815 (1960).

21. R. O. Litvinov and Hsü Tung-liang, this volume, p. 165.

22. J. R. Schrieffer, Semiconductor Surface Physics (New York, 1957), p. 55; F. S. Ham and D. C. Mattis, IBM J. Research Develop. 4, No. 2 (1960).

23. J. N. Zemel and R. L. Petritz, Phys. Rev. 110, 1263 (1958).

EFFECTIVE SURFACE RECOMBINATION
VELOCITY AND CRITERIA FOR ITS VALIDITY

V. A. Petrusevich and O. V. Sorokin

Institute for Semiconductors, Academy of Sciences, USSR

At present there are sufficient grounds for assuming that recombination at the surface plays a very important role in various phenomena connected with the existence of nonequilibrium carriers. Usually in solving the equation of motion for such carriers the surface recombination is allowed for by introducing into the boundary conditions a phenomenological parameter, the effective surface recombination velocity s_{eff}, and it is assumed that this parameter is independent of the way in which nonequilibrium carriers are generated, i.e., it is a constant for a given sample.

The use of s_{eff} has been found to be very fruitful both from the purely scientific and application points of view. Employing s_{eff}, it has been possible, on the one hand, to explain several interesting features in electronic processes in semiconductors, and, on the other, to determine the influence of the surface on the properties of semiconducting devices.

However, like every other phenomenological quantity, the effective surface-recombination velocity has restricted meaning and to understand correctly the physics of electronic processes in semiconductors and semiconducting devices, it is necessary to know precisely the limits of applicability of the concept of surface recombination velocity. This problem has been considered theoretically in [1, 2].

The purpose of the present work was to check experimentally the theoretical conclusions of Bir [2] in the case of germanium and silicon, to refine further the criteria of applicability of the concept of effective surface-recombination velocity, and to determine the influence of the geometry of the surface space charge on the spectral distribution of the photoconductivity (PC) and the photomagnetic effect (PME), which are frequently used to determine the magnitude of the surface recombination velocity.

1. Criteria for the Validity of the Effective Surface-Recombination Velocity

The analytical expression obtained by Bir [2] for the criteria of the validity of the concept of s_{eff} can be represented in the following form:

$$\left| \frac{1}{L} j_1 + \frac{1}{L^2} j_2 + \frac{s}{D} j_1 \right| \ll 1, \tag{1}$$

where

$$j_1 = L_D \int_1^{\psi_0} \frac{e^{\psi(x)} d\psi}{F}, \tag{2}$$

$$j_2 = L_D^2 \left(\int\limits_1^{\psi_0} \frac{e^{\psi(x)} d\Psi}{F} \right) \left(\int\limits_1^{\psi_0} \frac{e^{-\psi(x)} d\psi}{F} \right). \tag{3}$$

Here

$$F = [\gamma (e^{-\psi} - 1) + \gamma^{-1} (e^{\psi} - 1) + \psi (\gamma - \gamma^{-1})]^{\frac{1}{2}} \tag{4}$$

and

$$L_D = \left(\frac{\varepsilon kT}{8\pi e^2 n_i} \right)^{\frac{1}{2}}. \tag{5}$$

In the expressions (1)-(5) L is the diffusion length; D is the diffusion coefficient; $s = s_0 \exp(-\psi_0)$, where s is the "true" surface recombination velocity [2, 3]; ψ_0 is the height of the potential barrier at the surface in units of kT; ε is the dielectric constant; n_0 is the density of majority carriers in the interior; n_i is the density of intrinsic carriers; $\gamma = n_0/n_i$; and the other symbols have their usual meaning. The inequality (1) was obtained on the assumption that the diffusion theory is valid and that the volume carrier lifetime τ is the same at all points in the sample, including the space-charge layer at the surface.

To check the inequality (1), we studied germanium of near-intrinsic conductivity, as well as n- and p-type silicon of resistivity ρ from 10 to 130 $\Omega \cdot$cm. The quantity ψ_0 was determined by means of the field effect, while s, L, and D were found by methods described in [4]. Since the results of the check on the expression (1), the technique of preparation of the samples and the necessary measurements have been given in our earlier work [3], we shall limit ourselves here to the analysis of the main results, which are as follows.

1. Treatment of near-intrinsic germanium with the usual etchants produces a surface potential barrier, the height of which does not exceed 6 kT; but it is known [5, 6] that inversion layers are easily formed in low-resistivity germanium samples and, therefore, for $\rho \approx 1$ $\Omega \cdot$cm the quantity ψ_0 should be greater than 11. If field-effect measurements are carried out on such samples, the maximum curvature of the bands may be greater than 15 kT, since the depth of modulation in such experiments may frequently reach ± 8 kT.

2. In the case of silicon, etching may produce even higher surface barriers (up to 33 kT).

Using these data we integrated numerically the expressions (2) and (3) for ψ_0 within the limits 1-16 for germanium and 1-35 for silicon, and plotted the appropriate graphs shown in Figs. 1 and 2. The details of these calculations have been described in [3]. The plots in Figs. 1 and 2 can be used to show that in the case of silicon, even at the usual values of the diffusion length and relatively low blocking barriers, the inequality (1) is not obeyed. For our samples ($\gamma \approx 1.5 \times 10^5$, L = 0.03 cm), we have $L^{-2}j_2 > 1$ even when $\psi_0 > 14$ and $L^{-2}j_2$ rises rapidly with the increase of ψ_0; the point of inversion of the sign of surface conductivity is reached at $\psi_0 = 23$ and, consequently, it is legitimate to introduce s_{eff} only for low blocking and antiblocking barriers. The higher the resistivity of a material, the lower the values of ψ_0 (other conditions being equal) at which s_{eff} loses its meaning. From Fig. 2 it follows that in the case of germanium with the usual values of γ, ψ_0, and L, the inequality (1) is always satisfied, i.e., it would seem legitimate to introduce s_{eff}. Actually, for $L \approx 0.1$ cm we have $L^{-1}j_1 < 0.1$ (antiblocking layer, $\psi_0 < 10$ and $\gamma = 2$) and $L^{-2}j_2 < 10^{-2}$ (inversion layer, $\psi_0 < 16$ and $\gamma = 180$).*

However, the criterion of Eq. (1) may not be obeyed. For example, if $L < 10^{-2}$ cm, $\psi_0 > 15$ and $\gamma = 180$ then $L^{-2}j_2 > 1$ (inversion layer) or if $\psi_0 > 9$, $\gamma = 2$ then $L^{-1}j_1 > 1$ (antiblocking layer). It is necessary to point out that these estimates are unreliable for an inversion layer since it is not clear whether the criterion (1) is applicable. The shorter the diffusion length, the smaller the values of ψ_0 at which the inequality (1) ceases to be obeyed. The very small values of L are characteristic only of germanium samples containing considerable amounts of some impurities and also of samples of low resistivity, and, therefore, from the results just given one might conclude that in studies of sufficiently pure germanium we can practically always use the concept of the effective surface-recombination velocity. However, several observations show that this is not quite true.

---

* The remaining terms in Eq. (1) are much smaller than the values of the parameters used here.

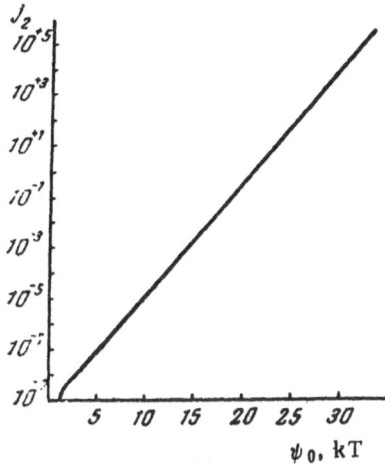

Fig. 1. Plot of the dependence of j_2 on ψ_0 for silicon. Blocking layer with $\gamma = 1.5 \times 10^5$.

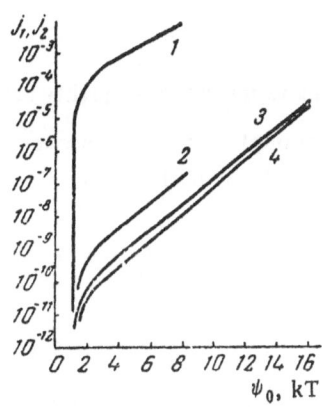

Fig. 2. Dependence of j_1 and j_2 on ψ_0 for germanium. 1) Antiblocking layer (j_1), $\gamma = 2$; 2) blocking layer (j_2), $\gamma = 10$; 3) blocking layer (j_2), $\gamma = 30$; 4) blocking layer (j_2), $\gamma = 180$.

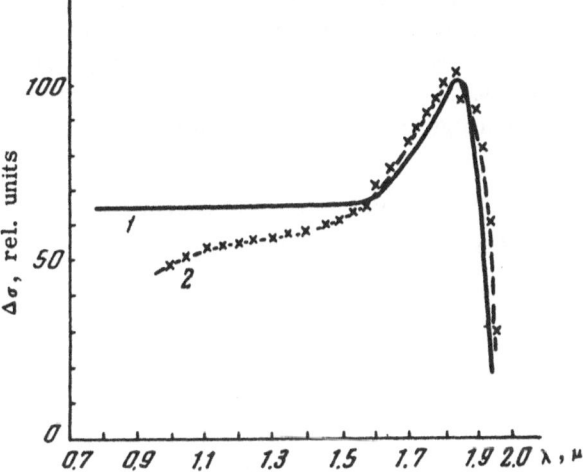

Fig. 3. Spectral distribution of the photoconductivity of p-type germanium ($\rho = 47$ $\Omega \cdot$cm, $d = 3.7$ mm). Continuous curve is theoretical for $s_{eff} = 10^3$ cm/sec; dashed curve is experimental for a sample etched in hydrogen peroxide.

(i) Comparing the results of measurements — made using the PME and PC methods — of the surface recombination velocity, diffusion length and ambipolar diffusion coefficient of germanium, it was found that when the volume parameters are measured these methods give very close values, while when s_{eff} is measured considerable discrepancies are frequently observed, and the higher the absolute value of s_{eff}, the higher the relative difference between the measured values of s_{eff} (for $s_{eff} \approx 10^4 - 10^5$ cm/sec these values differed by more than one order of magnitude).

(ii) The experimental PC curves did not always coincide with the theoretical curves calculated on the assumption that it is legitimate to introduce s_{eff}. Figure 3 is an example of such a case. Similar observations were also reported in [7, 8].

(iii) As pointed out by Rzhanov [9], the experimental dependences of s_{eff} on ψ cannot be explained if we assume the presence of one type of recombination center. It is possible that the discrepancies between the theoretical and experimental curves are due to some lack of rigor in the introduction of s_{eff}.

These and several other observations [3, 4] suggest that the criteria for the validity of the concept of s_{eff}, deduced by Bir [2], are insufficiently stringent and that in fact the region of applicability of s_{eff} is narrower.

Bir assumed that the volume carrier lifetime τ is the same at all points in the sample. However, if this assumption is not made, then solving the problem by the same method as used by Bir [2], we obtain the following expressions instead of (1) and (3):

$$\left| \frac{1}{L} j_1 + j_2' + \frac{s}{D} j_1 \right| \ll 1; \qquad (1')$$

117

$$j_2' = \frac{L^2}{D} \left(\int_1^{\psi_0} \frac{e^\psi \, d\psi}{F} \right) \left(\int_1^{\psi_0} \frac{e^{-\psi} d\psi}{\tau(x) \, F} \right). \tag{3'}$$

The remaining terms in the inequality (1) remain the same.

As an example, let us consider the case when the carrier recombination mechanism is described by the Shockley—Read theory [13]. For the sake of simplicity, let us assume that $\tau_n = \tau_p$, then

$$\tau = \tau_0 \left[1 + \frac{\cosh(E_i - E_t)}{\cosh(E_i - F \pm \psi)} \right]$$

or when $|E_i - F \pm \psi| \gg 1$

$$\tau(x) = \tau_0 \left[1 + A \exp\left(-|E_i - F \pm \psi| \right) \right], \tag{6}$$

where E_t, E_i, F are, respectively, the energies of the recombination centers, of the middle of the forbidden band and of the Fermi level in the interior, all expressed in units of kT.

Substituting Eq. (6) into Eq. (3'), we can easily see that, depending on the sign of ψ, the value of j_2' may be considerably larger or smaller than j_2, indicating that in some cases the criterion (1) is insufficiently stringent and that in these cases it is necessary to use the more exact expressions (1') and (3'). An analysis of the results of measurements using (1') and (3') meets with considerable difficulties because it is necessary to know the actual mechanism of volume recombination and the parameters on which the number of recombination acts depend (for example, in the case of the Shockley—Read recombination mechanism, it is necessary to know the density of recombination centers, their positions relative to the energy bands of the semiconductor and the cross sections for capture of electrons and holes).

Thus, it is clear that in some cases important in practice, for example, in the treatment of a silicon surface) such thick layers of space charge may form that it is not legitimate to introduce s_{eff}.

It has therefore become necessary to refine the theory of the photoconductivity* and the photomagnetic effect, for which the use of s_{eff} may be important, and which are frequently used [4] to determine the recombination constants and other parameters of semiconductors.

2. Photoconductivity

To make the case definite, we shall consider a rectangular plane-parallel sample of thickness d. We shall assume that both large surfaces of the plate have been subjected to identical treatment so that they are characterized by the same values of the true surface recombination velocity $\tilde{s}$ and the same surface potential barriers (blocking or antiblocking), the height of which varies linearly with the coordinate. This means that the electric field E is constant in the space-charge region. Figure 4 shows a one-dimensional band model representing this case. Let the intensity of light be so small that $\Delta n = n - n_0 \ll n_0$ and $\Delta p = p - p_0 \ll n_0$ and that the change in the electric field produced by illumination is negligibly small. For linear and homogeneous boundary conditions, sufficiently small E and small mean free path r, when the condition

$$\frac{q \tilde{l} E}{kT} \ll 1,$$

is satisfied, we can write the following system of equations [2]:

$$\left. \begin{aligned} \frac{1}{q} I_{1,3} &= \pm \mu E \Delta p_{1,3} - D \frac{d \Delta p_{1,3}}{dx} \, ; \\ \frac{1}{q} I_2 &= - D \frac{d \Delta p_2}{dx} \, ; \end{aligned} \right\} \tag{7}$$

$$\frac{1}{q} \frac{d I_{1,2,3}}{dx} + \frac{\Delta p_{1,2,3}}{\tau} = Q K e^{-kx}. \tag{8}$$

---

* The photoconductivity of an infinitely thick sample has been considered by Bir [2].

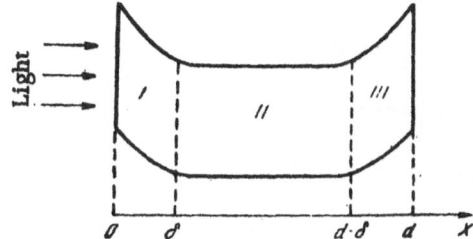

Fig. 4. Schematic representation of regions in the semiconductor for which the calculations are made.

Here, I is the nonequilibrium minority carrier current; μ and D are, respectively, the mobility and the diffusion coefficient of these carriers; Q is the density of the photon flux; K is the absorption coefficient. The subscripts "1," "2," and "3" indicate that the particular quantity corresponds to one of the regions I, II, III in Fig. 4.

If the quantities μ, D, and τ are independent of the coordinate x and if the electron and hole lifetimes are equal ($\tau_n = \tau_p = \tau$), then to determine the density of the nonequilibrium carriers in the sample, it is sufficient to solve Eqs. (7) and (8) under the following boundary conditions:

$$\frac{1}{q} I_1(0) = -\bar{s}\Delta p_1(0);$$
$$\frac{1}{q} I_3(d) = \bar{s}\Delta p_3(d). \tag{9}$$

Moreover, the following relationships should be satisfied:

$$\begin{aligned}
\Delta p_1(\delta) &= \Delta p_2(\delta), \\
I_1(\delta) &= I_2(\delta), \\
\Delta p_3(d-\delta) &= \Delta p_2(d-\delta), \\
I_3(d-\delta) &= I_2(d-\delta),
\end{aligned} \tag{10}$$

where δ is the thickness of the surface barrier [3].

In the steady-state, the photoconductivity $\Delta\sigma$ is given by the following expression:

$$\Delta\sigma = \frac{\beta q(\mu_n + \mu_p)}{d} \int_0^d \Delta p\, dx = A \int_0^d \Delta p\, dx, \tag{11}$$

where β is the quantum yield. Using relationships (7)-(11) we obtain

$$\begin{aligned}
\Delta\sigma = \frac{A}{K} &\left[C_1(1-e^{-K\delta}) + C_2 e^{-K\delta}(1-e^{-Kd'}) + C_3 e^{-Kd}(e^{K\delta}-1)\right] + \\
&+ \frac{C_1 AL}{B}\left\{(K+2\xi)\left[D(1-e^{-K\delta})F - G\bar{s}Le^{-K\delta}\right] + \right. \\
&\left. + \bar{s}L(\alpha_1+\alpha_2)e^{-K\delta} + H\right\} - \\
&- \frac{C_2 AL}{B} e^{-K\delta}(1-e^{-Kd'})\left\{\bar{s}(\alpha_1+\alpha_2)\coth\frac{Kd'}{2} + \right. \\
&\left. + K\left[B - \bar{s}L(\alpha_1+\alpha_2)\coth\frac{d'}{2L}\right]\right\} + \\
&+ \frac{C_3 ALe^{-Kd}}{B}\left\{(K-2\xi)\left[D(e^{K\delta}-1)F - G\bar{s}Le^{K\delta}\right] + \right. \\
&\left. + \bar{s}L(\alpha_1+\alpha_2)e^{K\delta} + H\right\}.
\end{aligned} \tag{12}$$

Here

$$B = (a_1 D + \bar{s})\left(1 + a_2 L \coth\frac{d'}{2L}\right)e^{a_1\delta} +$$

$$+ (a_2 D - \bar{s})\left(1 - a_1 L \coth\frac{d'}{2L}\right)e^{-a_2\delta};$$

$$F = (e^{-a_2\delta} - e^{a_1\delta})\coth\frac{d'}{2L} - a_1 L e^{a_1\delta} - a_2 L e^{-a_2\delta};$$

$$G = [a_1 L (1 - e^{-a_2\delta}) + a_2 L (1 - e^{a_1\delta})]\coth\frac{d'}{2L} - e^{a_1\delta} + e^{-a_2\delta};$$

$$H = \bar{s}(e^{-a_2\delta} - e^{a_1\delta})\coth\frac{d'}{2L} - \bar{s}L(a_1 e^{a_1\delta} + a_2 e^{-a_2\delta});$$

$$C_1 = \frac{QK\tau}{1 - K^2 L^2 - 2\xi KL^2}; \quad C_2 = \frac{QK\tau}{1 - K^2 L^2}; \quad C_3 = \frac{QK\tau}{1 - K^2 L^2 + 2\xi KL^2};$$

$$\xi = \frac{\mu E}{2D}; \quad a_1 = \left(\xi^2 + \frac{1}{L^2}\right)^{\frac{1}{2}} + \xi; \quad a_2 = \left(\xi^2 + \frac{1}{L^2}\right)^{\frac{1}{2}} - \xi;$$

$$d' = d - 2\delta.$$

It can be easily shown that: a) when $\xi = 0$ or $\delta = 0$, the expression (12) reduces to the formula derived without allowance for the geometry of the surface barrier [7, 10]; b) when $d = \infty$, the expression (12) reduces to the formula derived by Bir [2].

Several methods for the determination of the semiconductor parameters from the spectral distribution of the photoconductivity [11, 12] are based on the fact that when

$$KL \gg 1 \text{ and } Kd \gg 1 \tag{13}$$

$$\Delta\sigma = \text{const} \cdot \left(1 + \frac{s_{eff}}{D} \cdot \frac{1}{K}\right). \tag{14}$$

Let us consider the case of a thick sample ($d'/2L \gg 1$). When the inequalities (13) are satisfied, it follows from Eq. (12) that

$$\Delta\sigma = \frac{AQL}{B}\left[\frac{\psi_0 L}{\delta} - e^{-\psi_0} - \frac{s_{eff}L}{D}\left(\frac{\delta}{\psi_0 L}e^{\psi_0} + e^{-K\delta} + \frac{1}{KL} - \frac{\psi_0\left(e^{\psi_0} + \frac{\psi_0}{K\delta}\cdot e^{-K\delta}\right)}{K\delta + \psi_0}\right)\right]. \tag{15}$$

This expression can be reduced to (14) only for special values of the parameters occurring in it. In the general case when a thick space-charge layer is present at the surface, it is not possible to determine s_{eff} by the photoconductivity or any other method.

3. Photomagnetic Effect (PME)

From the standpoint of the use of the PME for the determination of the recombination constants, the cases of thick $d/L \gg 1$ and thin $d/L \ll 1$ samples are of practical interest. Moreover, in this case, the strong-absorption approximation ($K = \infty$) is frequently used, which in practice corresponds to the condition (13).

An analysis shows that in the presence of a thick surface barrier Eq. (13) should be rewritten as:

$$K\delta \gg 1 \text{ and } KL \gg 1 \tag{13'}$$

and moreover the inequality

$$\frac{2|\xi|}{K} \ll 1. \tag{13''}$$

should be satisfied.

The simplest expressions for the open-circuit photomagnetic emf, V_{pm}, are obtained for the condition of saturation of the PME at high illumination intensities or under the condition of the compensation of the PME by the photoconductivity [4]:

$$V_{pm} \sim \frac{\int_0^d I \, dx}{\int_0^d \Delta p \, dx} \, .$$

The expression obtained in this way is, in general, very complex and therefore we shall limit ourselves to the special case of a thick sample. When the conditions (13') are satisfied we obtain

$$V_{pm} = h \frac{BD}{10^8 L} \frac{\frac{\delta}{L} + \alpha_2^2 L^2 e^{-\alpha_1 \delta}}{1 + \alpha_2^2 L^2 e^{-\alpha_1 \delta}} \cdot \frac{1 + \frac{\delta}{\psi_0 L} e^{-\alpha_1 \delta} + \frac{\bar{s}}{\alpha_1 D}(1 + \alpha_1 L e^{-\alpha_1 \delta})}{1 - \frac{\delta}{\psi_0 L} e^{-\alpha_1 \delta} + \frac{\bar{s}}{\alpha_1 D}(1 - e^{-\alpha_1 \delta})} \, , \tag{16}$$

where h is the length of the illuminated portion of the sample, B is the magnetic field intensity.

When $\xi > 0$

$$V_{pm} = h \frac{B}{10^8} \frac{D}{L} \cdot \frac{\delta}{L} \, , \tag{17}$$

i.e., we obtain a value δ/L times smaller than that which follows from the usual theory [4].

When $\xi < 0$

$$V_{pm} = h \frac{B}{10^8} \frac{D}{L} \cdot \frac{\delta}{L} \frac{1 + \frac{\delta}{\psi_0^2 L} e^{\psi_0}}{1 + \frac{\delta^2}{\psi_0^2 L^2} e^{\psi_0}} \cdot \frac{1 - \frac{\delta}{\psi_0 L} e^{\psi_0} + \frac{s_{eff} L}{D}}{1 + \frac{\delta}{\psi_0 L} e^{\psi_0} + \frac{s_{eff} L}{D}} \, . \tag{18}$$

If $(\delta/\psi_0 L) e^{\psi_0} \ll 1$ then Eq. (18) reduces to Eq. (17), but if $(\delta/\psi_0 L)^2 e^{\psi_0} \gg 1$, then

$$V_{pm} = h \frac{B}{10^8} \frac{D}{L} \cdot \frac{\frac{s_{eff} L}{D} - \frac{\delta}{\psi_0 L} e^{\psi_0}}{\frac{s_{eff} L}{D} + \frac{\delta}{\psi_0 L} e^{\psi_0}} \, , \tag{19}$$

i.e., again the value of V_{pm} is low and for some ratios of the parameters V_{pm} may vanish altogether. It is possible that this explains the surprisingly small value of the photomagnetic effect observed by the present authors in CdTe single crystals. From the expressions (16)-(19), it is clear that when a thick layer of space charge is present at the surface, the diffusion length measured by means of the PME may be considerably in error. Similarly we may show that at high values of ψ_0 and δ, measurement of the surface recombination velocity (for this we should use the thin-sample approximation) by means of the PME also becomes practically impossible since the appropriate expressions are very complex and they contain surface barrier parameters.

Concluding, the authors express their deep gratitude to G. E. Pikus and G. L. Bir for their interest in this work and valuable advice.

LITERATURE CITED

1. F. Berz, Proc. Phys. Soc. (London) B71, 275 (1958).
2. G. L. Bir, Fiz. Tverd. Tela 1, 67 (1959).
3. V. A. Petrusevich, O. V. Sorokin, and V. I. Kruglov, Fiz. Tverd. Tela 3, 2023 (1961).
4. V. A. Petrusevich, V. K. Subashiev, and G. P. Morozov, Fiz. Tverd. Tela 3, 1505 (1961).
5. M. Lasser, C. Wysocki, and B. Bernstein, in the collection: Semiconductor Surface Physics [Russian translation] (IL, 1959), p. 247.
6. A. Many, E. Harnik, and G. Margoninski, in the collection: Semiconductor Surface Physics [Russian translation](IL, 1959), p. 127.
7. V. A. Petrusevich, Fiz. Tverd. Tela 1, 56 (1959).
8. Tang Ting-yuan, Acta Physica Sinica 13, No. 5, 427 (1957).
9. A. V. Rzhanov, this volume, p. 70.
10. H. B. De Vore, Phys. Rev. 102, 86 (1956).
11. V. K. Subashiev, V. A. Petrusevich, and G. B. Dubrovskii, Fiz. Tverd. Tela 2, 1022 (1960).
12. V. A. Petrusevich, Fiz. Tverd. Tela 3, 1268 (1961).
13. W. Shockley and T. V. Read, Phys. Rev. 87, 835 (1952).

DEPENDENCE OF THE SURFACE RECOMBI-
NATION VELOCITY OF GERMANIUM ON THE
MAJORITY CARRIER DENSITY

V. I. Strikha

T. G. Shevchenko State University, Kiev

The general expression obtained [1-3] for the dependence of the surface recombination velocity on the surface potential φ_s and on the density of majority carriers n_0 or p_0 in the case of recombination at levels of one type (the levels are at the energy position $E_t - E_i$, have cross sections for the capture of holes C_p and electrons C_n and are present in a density N_t) can be represented in the following way:

$$s = \frac{N_t C_p C_n (n_0 + p_0)}{n_i \left\{ C_n \exp \dfrac{E_t - E_i}{kT} + C_p \exp \dfrac{E_i - E_t}{kT} + C_n \exp \dfrac{e\varphi_s}{kT} + C_p \exp - \dfrac{e\varphi_s}{kT} \right\}} \cdot \tag{1}$$

where n_i is the equilibrium carrier density in an intrinsic semiconductor. This shows that the surface recombination velocity depends on many parameters.

While the dependence of the surface recombination velocity on the surface potential and the surface-level parameters has been investigated quite thoroughly [3-8, 11], the dependence of this velocity on the majority-carrier density has attracted very little attention. Only one paper is known [9] which reports a study of the change – within a narrow range – in the carrier density, for the same surface treatment.

It is of interest to investigate the dependence of the surface recombination velocity on the majority-carrier density in a wide range of the density values and for different surface treatments, because this gives:

1) the experimental dependence of the surface recombination velocity on the carrier density, suitable for practical purposes;

2) the factors governing the form of the dependence on the density, in particular the nature of the change in the surface recombination levels with the density.

The surface recombination velocity was measured on n-type germanium doped with antimony, and on p-type germanium doped with gallium, up to carrier densities of 10^{17} cm^{-3}. The measurements were made by modulating the conductivity [10] at two sample thicknesses, which made it possible to determine separately the surface recombination velocity and the volume lifetime of carriers. To reduce the error due to the scatter of the surface recombination velocity values, the surface treatment was carried out several times for the same sample thickness and the results were averaged out. The experiment was conducted in air on a freshly treated surface which had been dried in a stream of hot air.

The duration of etching, necessary for complete removal of the results of previous grinding, was established in separate tests.

The experimental dependence of the surface recombination velocity for a surface etched in boiling Perhydrol is given in Fig. 1 and that for a surface etched in CP-8 is given in Fig. 2. The surface recombination velocity was about an order of magnitude greater after etching in CP-8 compared with the velocity after etching in Perhydrol, and in both cases the velocity rose with increasing majority-carrier density both in n- and p-type germanium. A similar increase of the surface recombination velocity was obtained for a ground surface. But here the velocity was several times greater than that of a surface etched in CP-8. An analysis of the dependence of the surface recombination velocity on the carrier density was carried out only for etched surfaces since the ground surface had a highly defective layer where recombination could not be neglected.

It follows from the expression quoted above that the surface recombination velocity should be proportional to the majority-carrier density when C_p, C_n, N_t, φ_s, and $E_t - E_i$ are all constant. However, the figures show that the dependence of the surface recombination velocity on the carrier density is sublinear. What are the reasons for this observation?

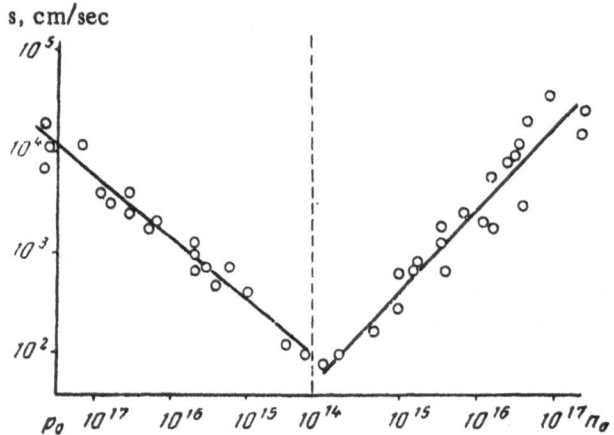

Fig. 1. Dependence of s on the majority-carrier density after etching in Perhydrol.

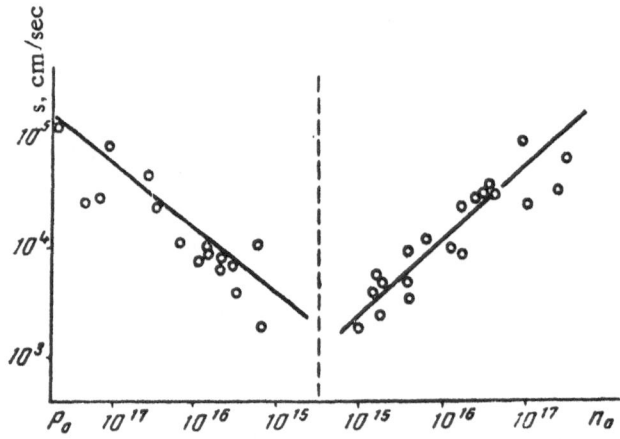

Fig. 2. Dependence of s on the majority-carrier density after etching in CP-8.

First, let us establish whether it is correct to use the surface recombination velocity in the form of Eq. (1), doubts about which were reported earlier [12, 13]. For this purpose, let us consider the validity of the diffusion theory in the case of our samples. Calculations showed that, at least for the low-resistivity samples, the diffusion theory is inapplicable and the restrictions mentioned in [12] do not apply.

Another possible reason was the inconstancy of the parameters in Eq. (1). Most probably N_t or φ_s are not constant. In fact, when the impurity concentration in the interior of germanium is greater than 10^{15} cm^{-3}, this concentration in the surface layer exceeds 10^{10} cm^{-2}, which is comparable with or greater than the surface level density ($10^{10} - 10^{12}$ cm^{-2}) in pure germanium. This may affect the surface states present earlier. However, when the impurity concentration in the interior is altered, even if the density of surface levels is not affected, there should be a change of φ_s. The lower the surface level density, the greater should be the change of φ_s.

It seems less likely that C_p, C_n, and E_t - E_i vary with the majority-carrier density.

To find the factor which governs the observed dependence of the surface recombination velocity, we examined the variation of this velocity with temperature. Let us consider the form of such a variation. We shall start from the assumption that recombination occurs mainly at one level. Such an assumption is in agreement with most of the published work [3, 5-8]. In that case, when one of the exponentials in the denominator is the dominant one, which can be true in a narrow range of temperature, Eq. (1) is transformed into

$$s = \frac{N_t C_{p,n} (n_0 + p_0)}{\sqrt{N_c N_v} \, \exp\left(\pm \frac{\Delta E}{kT} \right)}, \tag{2}$$

where $\pm\Delta E$ is equal to the sum of the numerators in the powers of the exponentials representing the temperature dependences of n_i, φ_s, or E_t - E_i, and possible C_p or C_n. Here it is assumed that φ_s is independent of temperature. Control experiments on the temperature dependence of the contact potential difference confirmed the correctness of this assumption. Thus when Eq. (2) is valid, the dependence $\ln sT = f(1/T)$ should be a straight line, from the slope of which we can find $\pm\Delta E$.

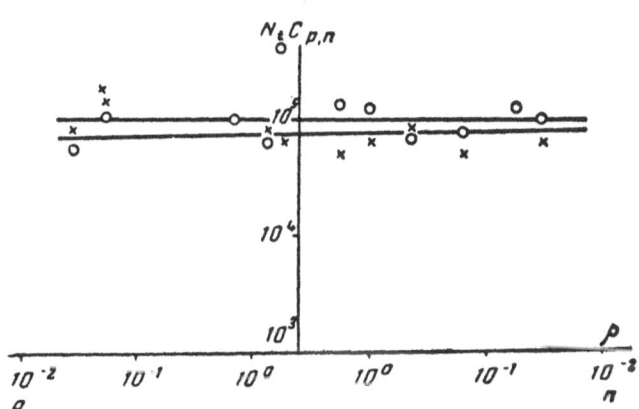

Fig. 3. Dependence of the product of the surface-level density and the effective capture cross section on the resistivity of germanium: × - etched in Perhydrol; ○ - etched in CP-8.

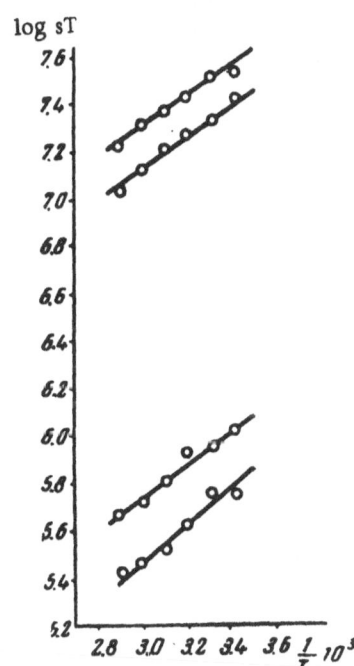

Fig. 4. Dependence of the surface recombination velocity on temperature.

The measurements showed that for all samples the dependence $\ln sT = f(1/T)$ can be satisfactorily represented by straight lines in the range 290-360°K. Dependences for four samples of n-type germanium etched in CP-8 are shown in Fig. 4.

The values of $\pm\Delta E$ so determined allowed us to calculate the product $N_t C_{p,n}$ for samples with different initial resistivities ρ. The results of the calculations are shown graphically in Fig. 3. It is seen that, in spite of considerable scatter, the product $N_t C_{p,n}$ is approximately constant, or at least there is no marked tendency to increase or decrease. This, more than anything else, supports the hypothesis that the surface-level density is independent of the impurity concentration in the interior.

Moreover, it should be noted that the product $N_t C_{p,n}$ is approximately the same after surface treatment in CP-8 or Perhydrol. At the same time, the surface recombination velocities after these treatments differ by an order of magnitude, which is related to differences in other parameters which occur in Eq. (1).

All this does not contradict the hypothesis that recombination in the test samples occurs at levels having energy positions at about +4 kT and densities of 5×10^{11} cm^{-2}. These levels appear clearly in a study of the change in the conductivity in an external field.

LITERATURE CITED

1. W. Brattain and J. Bardeen, Bell System Tech. J. 32, 1 (1953).
2. A. Stevenson and R. Keys, Physica 20, 1041 (1954).
3. A. Many, E. Harnik, and Y. Margoninski, Semiconductor Surface Physics (New York, 1957), p. 85.
4. H. K. Henisch, P. W. Reynolds, and P. M. Tipple, Physica 20, 1033 (1954).
5. H. K. Henisch and P. W. Reynolds, Proc. Phys. Soc. (London) B66, 853 (1955).
6. A. Many and D. Gerlich, Phys. Rev. 107, 404 (1957).
7. A. V. Rzhanov, N. M. Pavlov, and M. A. Selezneva, Zhur. Tekh. Fiz. 28, 2645 (1958).
8. V. I. Lyashenko, O. V. Snitko, and T. N. Sytenko, Fiz. Tverd. Tela 1, 11 (1959).
9. B. Schultz, Physica 20, 1031 (1954).
10. K. D. Glinchuk, E. G. Miselyuk, and É. I. Rashba, Zhur. Tekh. Fiz. 26, 2607 (1956).
11. V. Ya. Priimachenko, V. G. Litovchenko, V. I. Lyashenko, and O. V. Snitko, Ukrain. Fiz. Zhur. 5, 346 (1960).
12. G. L. Bir, Fiz. Tverd. Tela 1, 67 (1959).
13. V. A. Petrusevich, O. V. Sorokin, and V. I. Kruglov, Fiz. Tverd. Tela 3, 2023 (1961).

FREE-RADICAL RECOMBINATION PROCESSES ON SEMICONDUCTOR SURFACES AND THEIR ROLE IN LUMINESCENCE

F. F. Vol'kenshtein, A. N. Gorban', and
V. A. Sokolov

Tomsk Polytechnical Institute; Physical Chemistry Institute,
Academy of Sciences, USSR

Back in 1925, Bonhoeffer [1], investigating hydrogen reactions, discovered the luminescence of crystal phosphors under the action of atomic hydrogen. Subsequently, other workers observed the luminescence of phosphors under the action of free atoms and radicals of other gases [2, 3]. The source of excitation of the luminescence in these cases is the energy evolved during the recombination of free atoms and gas radicals on the crystal surface. The crystal then acts as a catalyzer, without which the recombination reaction is unlikely (such a reaction in gas usually takes place on triple collision, when the recombining atoms and radicals can transfer the energy evolved to the third member).

Unfortunately, the very interesting cases (from the point of view of catalysis) of radical-recombination luminescence have been forgotten for many years and not studied specifically.

Recently, this form of luminescence has attracted attention in connection with investigations of the phenomenon of candoluminescence, i.e., luminescence of crystal phosphors under the action of chemically active flames. The history of candoluminescence studies is quite long and involved. Some workers accepted the existence of candoluminescence using various hypotheses to explain it (the oxidation-reduction hypothesis was most widely used). Others rejected the existence of candoluminescence as a special phenomenon and explained the observed features of radiation by purely thermal radiation in flames [4].

Now we may regard it as proved that the luminescence of crystal phosphors under the action of chemically active flames does indeed take place at temperatures below the quenching temperature of luminescence and represents a special type of true luminescence in the sense of the Vavilov-Wiedemann definition [5].

N. A. Prilezhaeva suggested, and Sokolov with Gorban' proved experimentally [6, 7], that the luminescence of crystal phosphors in flames and in atmospheres of active gases is due to processes of recombination, on the phosphor surface, of free atoms and radicals present in the flame and responsible for the chain reaction of combustion. This was proved by the following experiments:

1) observation of luminescence in the gas between the flame cones when other excitations (corpuscular, ultraviolet radiation of the flames, etc.), apart from radical recombination, were absent;

2) quenching of candoluminescence by the introduction, into a city-gas flame, of a copper grid in front of the phosphor (such a grid is a catalyzer for reactions of recombination of atomic hydrogen and other radicals) which weakens the recombination on the phosphor surface;

3) establishment of the identity of the luminescence spectrum obtained by a gas discharge and the candoluminescence spectrum in the flame of the same gas;

4) determination of the coefficients of recombination of atoms and radicals on some candoluminescence phosphors directly in flames.

In the present work we use the electronic theory of chemisorption and catalysis [8] to discuss the radical-recombination mechanism of luminescence and some consequences of this. The influence of an external transverse field on the intensity of candoluminescence was investigated and it was found that the experimentally determined dependences are in qualitative agreement with those expected from theory.

1. Radical-Recombination Mechanism of Luminescence

The process of luminescence consists usually of two consecutive stages. The first is the ionization and the second the neutralization of an activator atom.

To make the case definite, we shall assume that the activator atoms are donors, i.e., they are represented by donor local levels in the energy spectrum of the crystal. The treatment can be extended easily to the case of acceptor-type activator atoms. We shall denote a neutral atom of the activator by AL; eL and pL represent, respectively, a free electron and a free hole in the lattice; ApL is an ionized activator atom (an activator atom with a hole localized at it); and L denotes the lattice.

The activator may be ionized in the following two ways:

$$AL \rightarrow ApL + eL \tag{1}$$

or

$$AL + pL \rightarrow ApL. \tag{2}$$

In case (1) the ionization produces an electron in the conduction band, while in case (2) it annihilates a hole in the valence band. In the former case, the ionization requires energy, while in the latter case energy is evolved.

$$ApL + eL \rightarrow AL \tag{3}$$

or

$$ApL \rightarrow AL + pL. \tag{4}$$

In (3) the neutralization is accompanied by the emission of a quantum (luminescence); in (4) there is an energy loss and no luminescence. In this latter case, we are dealing with the transition of an electron from the valence band to the free level of the activator; the transition may be thermal in origin.

Thus luminescence may be represented by the mechanisms (1−3) or (2−3). The second of these requires first the formation of an electron−hole pair in the lattice, i.e., the ionization of the lattice

$$L \rightarrow eL + pL. \tag{5}$$

Figure 1 shows the energy spectrum of the crystal. The processes (1), (2), (3), (4), and (5) represent electron transitions, indicated by vertical arrows in Fig. 1.

In the case when luminescence proceeds according to the mechanism (1−3), the phosphor is excited as a result of ionization of the activator atoms (Fig. 1, transition 1); in the case of the (2−3) mechanism, the phosphor is excited by ionization of the lattice itself (transition 5 in Fig. 1).

The usual luminescence, excited by light quanta, is represented by the mechanism (1−3), while the candoluminescence is given by the mechanism (2−3). In the latter case, the phosphor is excited, i.e., electrons are transferred from the valence to the conduction band (transition 5) by the recombination of free atoms and radicals on the crystal surface.

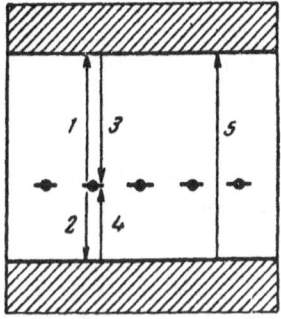

Fig. 1. Electron transitions during luminescence.

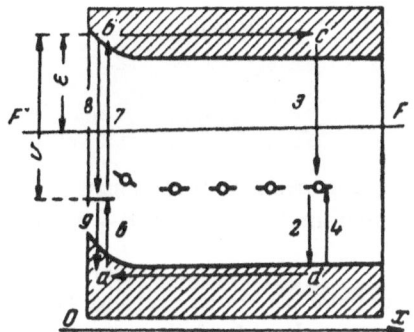

Fig. 2. Electron transitions in the radical-recombination mechanism of luminescence.

We shall assume that atoms or radicals taking part in the recombination (we shall denote them by the symbol C) are acceptors. The process of recombination giving rise to the formation of an electron–hole pair consists of the following consecutive stages (cf. Sec. 3a and 5a in [8]):

$$C + L \rightarrow CL;$$
$$CL \rightarrow CeL + pL; \tag{6}$$
$$C + CeL \rightarrow C_2 + eL. \tag{7}$$

The first of these represents the chemisorption of an atom or a radical C and the formation of a "weak" bond with the surface. The second converts the "weak" bond into a "strong" acceptor bond. This conversion is produced by a thermal transition of an electron from the valence band to the local acceptor level of the adsorbed particle and is accompanied by the formation of a free hole in the valence band. The third process represents the recombination of an atom or radical C, incident from the gas phase, with a chemisorbed atom or radical C strongly bound to the surface. This recombination is accompanied by the desorption of the molecule C_2 and the appearance of an electron in the conduction band. The transitions (6) and (7) are represented by arrows in Fig. 2.

Let us assume now that atoms or radicals C are not acceptors but donors. Then, instead of Eqs. (6) and (7), we shall have, respectively:

$$CL \rightarrow CpL + eL; \tag{6'}$$
$$C + CpL \rightarrow C_2 + pL. \tag{7'}$$

In the donor case, the act of "strengthening" the bond, represented by Eq. (6'), is accompanied by the appearance of a free electron and not of a hole, and the recombination stage (7') gives rise to a free hole and not a free electron. Irrespective of the acceptor or donor nature of the chemisorbed particles C, the sequence of acts of the bond "strengthening" and recombination produces a pair consisting of a free electron and a free hole.

Let us consider these processes in greater detail. We shall restrict ourselves to the case when particles C are acceptors. All that follows can be easily extended to the case of donor particles.

I. The act of "strengthening" the bond represents a transition of a chemisorbed acceptor particle from the electrically neutral state CL to the negative charged state CeL. This transition may proceed according to Eq. (6), as well as according to

$$CL + eL \rightarrow CeL \tag{8}$$

(transition 8 in Fig. 2). In the latter case, the "strengthening" of the bond of a given chemisorbed particle is due to electron transfer to the conduction band as a result of, for example, a recombination act of another chemisorbed particle, i.e., as a result of stage (7).

Along with the processes of bond "strengthening," there are the reverse processes of bond "weakening", according to the mechanism

$$CeL + pL \rightarrow CL$$

or

$$CeL \rightarrow CL + eL.$$

The former represents the recombination of an electron, localized at a chemisorbed particle, with a hole in the valence band; the latter represents a thermal transition of an electron to the conduction band, and we shall neglect it.

Thus there is electron exchange between a chemisorbed particle and the energy bands. The exchange with the valence band is of thermal origin, while that with the conduction band is due to the recombination (7).

II. The recombination of free atoms or radicals C on the surface of a phosphor may proceed according to Eq. (7) or according to the following equation

$$C + CL \rightarrow C_2 + L, \tag{9}$$

i.e., a chemisorbed atom (radical) can take part in recombination irrespective of whether it is in the state of "strong" or "weak" bonding [case (9)] with the surface. We note, however, that stage (9), in contrast to stage (7), is useless from the point of view of luminescence since it does not liberate an electron and leaves the phosphor unexcited. Thus not every recombination process may produce luminescence but only those proceeding according to Eq. (7) are effective.

This recombination mechanism was investigated quantum-mechanically in the case when the particle C is an atom with one valence electron [9]. It was shown that as the gas atom C approaches the chemisorbed atom CeL, a covalent bond of gradually increasing strength is formed between the two atoms while the binding of the chemisorbed atom to the lattice is weakened, the local level CL (Fig. 2) is drawn up to the conduction band (the distance v in Fig. 2 decreases) and the electron at this level is gradually delocalized. Consequently, after overcoming a certain activation barrier, the local level is drawn into the conduction band, the electron at this level becomes completely delocalized, and the bond between the chemisorbed atom C and the lattice is broken.

So far, we have considered the recombination of two atoms or radicals on the assumption that one of them is chemisorbed and the other is incident from the gas phase [cf. (7) and (9)]. We note that another situation is possible when both recombining atoms or radicals are in the chemisorbed state. In that case, instead of (7) and (9), we have

$$CL + CeL \rightarrow C_2 + eL \tag{7''}$$

and

$$CL + CL \rightarrow C_2 + L. \tag{9'}$$

From the point of view of luminescence it is unimportant whether recombination proceeds according to Eqs. (7) and (9) or according to Eqs. (7'') and (9') since both (7) and (7'') produce the same result: the appearance of an electron in the conduction band.

Thus, the radical-recombination luminescence may be represented as a cyclic process abcd in Fig. 2. The whole mechanism is started by an act of recombination which gives rise to transition (7) and has, as one of its stages, the phenomenon of luminescence represented by the transition (3).

2. Intensity of Luminescence

We shall now determine the factors which govern the intensity of luminescence produced by the radical-recombination mechanism.

We shall assume that the half-space $x \geq 0$ is occupied by a phosphor and that the half-space $x < 0$ is the gas phase (Fig. 2). The electron transitions shown in Fig. 2 are of two types: those in which local levels of the chemisorbed particles take part and those in which the activator levels are active (transitions 6, 7, 8, 9 and

2, 3, 4, respectively). The number of transitions of type i occurring in unit time on unit area will be denoted by s_i (here $i = 6, 7, 8, 9$). The quantity $r_k(x)$ denotes the number of transitions of the k-th type occurring in unit time in unit volume and referred to the x-plane (where $x \geq 0$, $k = 2, 3, 4$).

In the steady-state case, we have (for all $x \geq 0$):

$$s_6 - s_9 = s_7 - s_8,$$ (10)

$$r_2(x) - r_4(x) = r_3(x),$$ (11)

$$-\frac{dj_n}{dx} = -\frac{dj_p}{dx} = r_3(x),$$ (12)

where $j_n(x)$ and $j_p(x)$ are the electron and hole currents in the x-plane. It is clear that

$$j_n(x) = j_p(x),$$

and

$$\left. \begin{array}{l} j_n(\infty) = j_p(\infty) = 0 \\ j_n(0) = j_p(0) = j_s \end{array} \right\},$$ (13)

where

$$j_s = s_7 - s_8.$$ (14)

Integrating Eq. (12) over the whole volume of the crystal, we have, according to Eq. (13)

$$j_s = \int_0^\infty r_3(x)\, dx = I,$$ (15)

where I is the intensity of luminescence, defined as the number of quanta emitted by a unit area of the phosphor per unit time.

We shall subsequently assume that

$$s_8 \ll s_7,$$ (16)

where, from Eqs. (7) and (8),

$$s_7 = \alpha P N^-,$$ (17)

$$s_8 = \beta n N^0.$$ (18)

Here, P is the partial pressure of the gas consisting of particles C; N^- and N^0 are the numbers of such particles chemisorbed on a unit area of the surface and which are, respectively, in the charged and electrically neutral states, i.e., in the states of "strong" and "weak" binding with the surface (states CeL and CL); n is the density of conduction electrons in the plane $x = 0$; α and β are the coefficients of proportionality, which depend on temperature but are of no present interest.

According to Eqs. (15), (14), (16), and (17), we have

$$I = \alpha P N^-.$$ (19)

We shall assume that

$$\left. \begin{array}{l} s_7 \ll s_6 + s_9 \\ r_3 \ll r_2 + r_4 \end{array} \right\}$$ (20)

(the latter condition is assumed to be fulfilled for all $x \geq 0$). Then Eqs. (10) and (11) may be rewritten as

$$\left. \begin{array}{l} s_6 = s_9 \\ r_4 = r_2 \end{array} \right\},$$

and our problem can be considered as one of thermodynamic quasi-equilibrium. For N^- and N^0, we can use the formulas obtained in the electron theory of chemisorption for the case of electron equilibrium.

131

We have [cf. Eq. (24a) on p. 68 in [8]]

$$\eta^- = \frac{N^-}{N} = \left[1 + \exp\left(\frac{\varepsilon - v}{kT}\right)\right]^{-1} \Bigg\}$$
$$\eta^0 = \frac{N^0}{N} = \left[1 + \exp\left(\frac{v - \varepsilon}{kT}\right)\right]^{-1}$$

(21)

where

$$N = N^0 + N^-$$

is the total number of particles chemisorbed by a unit area of the surface; T is the absolute temperature; the meaning of the quantities ε and v in Eq. (21) is clear from Fig. 2, where the straight line FF represents the Fermi level. When the equilibrium between the surface and gas phase is established, we have [cf. Eqs. (27) and (29) on p. 70 in [8]]

$$N = N^* \left[1 + \eta^0 \frac{b}{P}\right]^{-1},$$

(22)

where N^* is the maximum number of particles which can be chemisorbed by a unit area (N^* is the number of particles chemisorbed on a unit area when $P = \infty$; obviously $N^* = 1/\sigma$, where σ is the area occupied by a chemisorbed particle); b is an adsorption coefficient which depends on temperature but its form is of no interest to us.

From Eqs. (19), (21), and (22) we obtain

$$I = \alpha P \eta^- N = \alpha N^* P \left[\frac{1}{\eta^-} + \frac{\eta^0}{\eta^-} \cdot \frac{b}{P}\right]^{-1}$$

or, after substituting Eq. (21):

$$I = A \left[1 + B \exp\left(\frac{\varepsilon - v}{kT}\right)\right]^{-1},$$

(23)

where

$$A = \alpha N^* P \Bigg\}$$
$$B = 1 + \frac{b}{P}$$

(24)

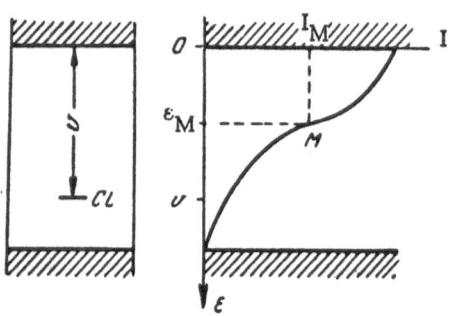

Fig. 3. Dependence of the luminescence intensity on the position of the Fermi level (the case of acceptor chemisorbed particles).

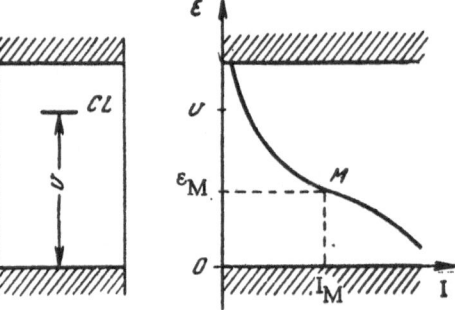

Fig. 4. Dependence of the luminescence intensity on the position of the Fermi level (the case of donor chemisorbed particles).

It is clear that at a given temperature and pressure (the values of P and T are fixed by the values of the parameters A and B) the intensity of luminescence is governed by the position of the Fermi level at the surface of the crystal (we mean here the Fermi level which represents the state of the system when the recombination and luminescence acts are neglected). The dependence of I on ε given by Eq. (23) is shown schematically in Fig. 3.

At the inflection point M (Fig. 3) on the I = I(ε) curve we have, according to Eq. (23):

$$\left.\begin{array}{l} \varepsilon_M = v - kT \ln\left(1 + b/P\right) \\ I_M = \frac{1}{2}\, aN^*P \end{array}\right\}. \tag{25}$$

From Eq. (25) it is clear that as P decreases, the inflection point in Fig. 3 is displaced to the right and upwards, disappearing altogether at

$$P \leqslant b\left[\exp\left(\frac{v}{\kappa T}\right) - 1\right]^{-1}.$$

As P increases, the inflection point shifts to the left and downwards, reaching the position $\varepsilon_M = v$ when $p = \infty$.

It follows, therefore, that the radical-recombination luminescence is favored by the condition

$$\varepsilon < \varepsilon_M. \tag{26}$$

Then, firstly, $\varepsilon_M > 0$, i.e., sufficient surface coverage with chemisorbed particles is ensured, and secondly $\varepsilon < v$, i.e., most of these particles are in the charged and not in the neutral state, and the recombination proceeds according to mechanism (7) and not mechanism (9), which does not produce luminescence.

It should be noted that in the derivation of Eq. (23), we assumed that the chemisorbed particles C taking part in recombination processes are acceptors. In the case of donor particles, it can be shown that we obtain the same formulas (23) and (26) in which, however, the energy separations ε and v are taken not from the bottom of the conduction band but from the top of the valence band, while the parameters A and B have their previous form given by Eq. (24). The dependence of I on ε in the donor case is shown schematically in Fig. 3, where ε_M and I_M are, as before, given by Eq. (25).

It is clear that the dependences of the luminescence intensity on the position of the Fermi level are opposite in the acceptor and donor cases: as the Fermi level is shifted downwards and other conditions are fixed, the luminescence is quenched in the case of acceptor particles but enhanced in the case of donor particles.

Equation (23) opens the way for the control of the intensity of the radical-recombination luminescence.

3. Influence of an External Electric Field on Luminescence

If the candoluminescence proceeds according to the radical-recombination mechanism we have described its intensity, as just shown, should be sensitive to factors which displace the Fermi level at the phosphor surface.

One such factor is foreign chemically inactive gas introduced into the flame, the molecules of which becoming adsorbed on the phosphor surface produce an additional surface charge. A foreign acceptor gas bends the energy bands upwards and consequently depresses the Fermi level. A donor gas bends the energy bands downwards and displaces the Fermi level upwards.

Thus the introduction of some foreign gas into a flame should, other conditions being unchanged, weaken or intensify the luminescence. It would be interesting to check this theoretical prediction experimentally.

Another factor is an external electric field applied to the phosphor. Let us consider a phosphor placed in a uniform electric field of intensity E directed at right angles to the adsorbing surface. The band curvature is altered and therefore the Fermi level at the surface is shifted. When E > 0 the Fermi level is shifted upwards, when E < 0 it is shifted downwards, as shown in Fig. 5, where $\varepsilon°$, ε^+, and ε^- denote the depth of the Fermi level below the conduction band in the absence of a field, for positive and negative directions of the field, respectively ($\varepsilon^+ < \varepsilon° < \varepsilon^-$).

Thus the application of a field should alter the intensity of luminescence. When the field direction is reversed, weakening of the luminescence should be replaced by enhancement, or conversely. This result is, in the final analysis, due to a change in the specific adsorption of the surface under the influence of the field, which is the effect investigated earlier [10] and following directly from the electron theory of chemisorption.

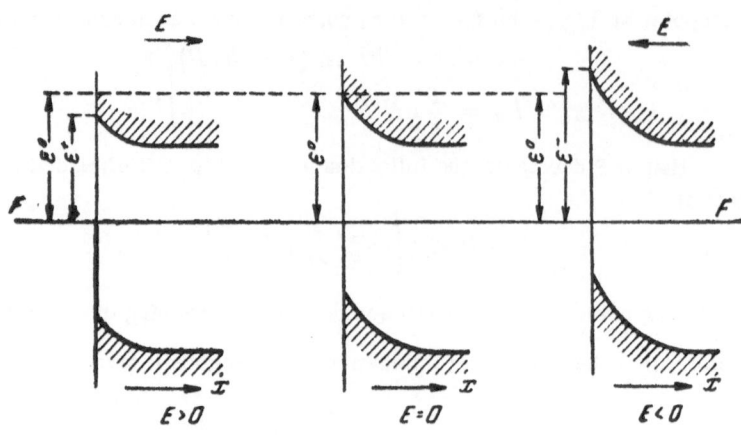

Fig. 5. Displacement of the Fermi level at the crystal surface under the
influence of an external electric field.

In connection with this theoretical prediction two of the present authors (Sokolov and Gorban') carried out experiments designed to detect the influence of an external electric field on the intensity of candolumi-nescence.

ZnS · CdS phosphor, activated with copper, was introduced into a low-temperature, low-pressure (8-20 mm Hg) flame of city gas. The phosphor was deposited on a glass substrate, the opposite side of which was coated with a conducting layer which served as an electrode. The apparatus, similar to that described earlier [11], is shown in Fig. 6. Here 1 is a mixer; 2 is a quartz reaction tube through which a stream of active gas was passed; 3 is a spiral for igniting the hot gas mixture; 4 is an electrode with a glass substrate and the phosphor deposited on it; 5 is a metal grid used as a second electrode; 6 is a buffer vessel. After the intensity of luminescence be-came constant an electric field was switched on. The observed change in the intensity of luminescence was recorded with a photomultiplier FEU-19M having a mirror galvanometer M-21 connected to its output.

The results of these experiments are given below. The first line gives the voltage V (in kilovolts) applied to the electrodes, the positive direction being (in accordance with Fig. 5) that which is shown by the vertical arrows in Fig. 6. The second line gives the intensity of luminescence I, in relative units (the intensity in the absence of a field being taken as unity).

V	−2	0	+2
I	1.14	1	0.25

The interpretation of the results within the framework of the ideas developed in this paper led to two conclusions.

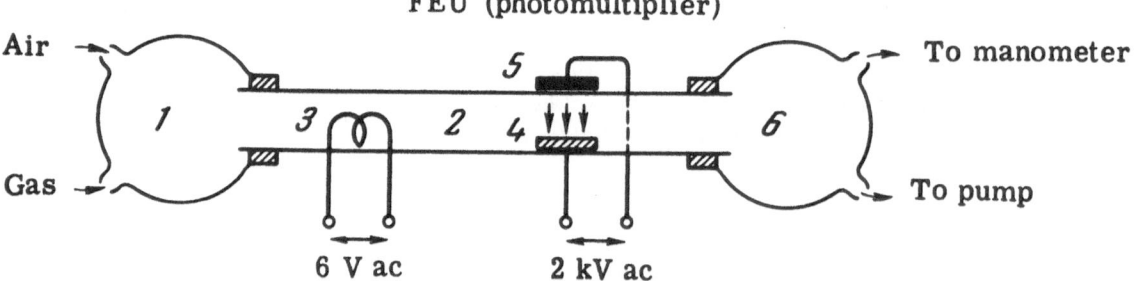

Fig. 6. Diagram of the apparatus used to study the influence of an external
electric field on the intensity of luminescence.

(i) The enhancement of the luminescence intensity on the application of a negative field (i.e., on downward displacement of the Fermi level) and its weakening by a positive field indicate that the chemisorbed particles, taking part in recombination (Fig. 4), are of the donor type. They may be hydrogen atoms present in the flame. In the case of acceptor particles we would have had the opposite result (Fig. 3).

This does not mean, however, that the flame gas (in fact a mixture of gases) charges the phosphor surface positively. It simply indicates that the particles which take part in recombination and produce luminescence contribute positive charge to the total surface charge.

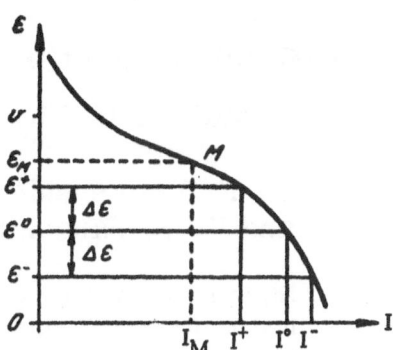

Fig. 7. Variation of the intensity of the radical-recombination luminescence under the influence of an external electric field.

(ii) The asymmetry of the effect, i.e., weak enhancement (by 14%) of the luminescence intensity by a negative field and pronounced weakening (by 75%) by a positive field of the same strength indicates that the Fermi level is initially (i.e., in the absence of a field) below the inflection point M in Fig. 4. If it had been above the point M, we would have had the opposite result. This is explained in Fig. 7 which repeats the information in Fig. 4 and in which ε°, ε^{+}, and ε^{-} denote the positions of the Fermi level in the absence of a field, a positive field and a negative field, respectively, while I°, I^{+}, and I^{-} represent the corresponding luminescence intensities. It is clear that

$$I^{-} - I^{\circ} < I^{\circ} - I^{+} \text{ when } \varepsilon^{\circ} < \varepsilon_{M},$$

$$I^{-} - I^{\circ} > I^{\circ} - I^{+} \text{ when } \varepsilon^{\circ} > \varepsilon_{M}.$$

Thus the experimental results indicate that the condition (26) is fulfilled, i.e., that the radical-recombination mechanism is probable under our conditions.

The results of the study of the influence of an electric field on the intensity of candoluminescence can be considered as one more experimental proof of its radical-recombination mechanism.

These results can be considered also as an indirect experimental confirmation of the theoretically expected effect of the influence of an electric field on the specific adsorption investigated earlier [10].

LITERATURE CITED

1. K. F. Bonhoeffer, Z. physik. Chem. **116**, 391 (1925).
2. H. S. Taylor and G. I. Lavin, J. Am. Chem. Soc. **52**, 1910 (1930).
3. P. N. Kokhanenko, Zhur. Fiz. Khim. **12**, 131 (1938).
4. V. A. Sokolov, Uspekhi Fiz. Nauk **47**, 537 (1952).
5. V. A. Sokolov, Optika i Spektroskopiya **4**, 409 (1958).
6. A. N. Gorban' and V. A. Sokolov, Optika i Spektroskopiya **7**, 259 (1959).
7. A. N. Gorban' and V. A. Sokolov, Izvest. Akad. Nauk SSSR, Ser. Fiz. **25**, 424 (1961).
8. F. F. Vol'kenshtein, Electron Theory of Catalysis on Semiconductors (Fizmatgiz, 1960).
9. F. F. Vol'kenshtein, Izvest. Akad. Nauk SSSR, Otdel. Khim. Nauk 143 (1957).
10. F. F. Vol'kenshtein and V. B. Sandomirskii, Doklady Akad. Nauk SSSR **118**, 980 (1958).
11. V. N. Panfilov, Yu. D. Tsvetkov, and V. V. Voevodskii, Kinetika i Kataliz **1**, 333 (1960).

CHEMISORPTIVE AND CATALYTIC PROPERTIES
OF A SEMICONDUCTING LAYER ON A METAL

F. F. Vol'kenshtein, V. S. Kuznetsov,
and V. B. Sandomirskii

Physical Chemistry Institute, Academy of Sciences, USSR;
Institute for Radio Electronics, Academy of Sciences, USSR;
Institute for Catalysis, Academy of Sciences, USSR

In most circumstances, the majority of metals are covered with a layer of a binary compound. i.e., they are coated with a semiconducting sheath. Thus the chemisorptive and the catalytic processes that we regard as occurring on a metal surface in fact take place on the surface of its semiconductor coating (this has been frequently stressed by many authors, cf. for example [1, 2]) and can therefore be considered within the framework of the electron theory of chemisorption and catalysis on semiconductors.

When one speaks of the catalytic action of a metal, it is frequently catalytic action of the semiconducting sheath that is in question but with the metal properties affecting the catalysis. In fact, if the thickness of the semiconductor layer is small compared with the screening length, then, according to the electron theory of catalysis, one should expect the chemisorptive and catalytic properties of the layer to depend on its thickness and on the properties of the metal underneath it.

The purpose of the present work was to investigate these dependences. Since the specific adsorption and the catalytic activity are governed, according to the electron theory, by the position of the Fermi level at the surface of the crystal, the problem is reduced to the determination of the Fermi level position at the outer surface of the layer as a function of the layer thickness L and of parameters representing the nature of the metal. In the present study, the work function of the metal χ was used as such a parameter.

1. Statement of the Problem

We shall consider a metal covered by a plane-parallel layer of uniform semiconductor, containing, in general, both donor and acceptor impurities distributed uniformly in the layer. Gas is chemisorbed on the free surface of the semiconductor. We shall assume that the surface states at the metal — semiconductor boundary are altogether absent and that the outer surface of the semiconductor has surface states both of adsorptive and nonadsorptive origin, which are responsible for the surface charge σ_L on the outer surface of the layer. We shall further assume that the positions of the surface levels are independent of the layer thickness L. As a result of contact with the semiconducting layer, the metal acquires a charge σ_0.

Figure 1 shows the energy diagram for a metal covered by a fairly thick semiconducting layer which does not have any surface states at all (the case $L \gg l$, where l is the screening length; $\sigma_L = 0$). Here φ is the photoelectric work function and ε_0 is the thermionic work function of the semiconductor. It is clear that

$$\chi = \varepsilon_0 + \zeta,$$

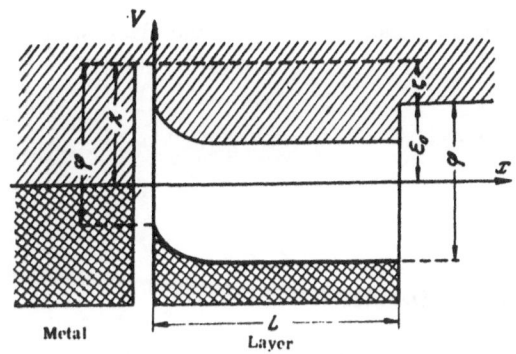

Fig. 1. Energy-band model for a metal covered
with a thick semiconductor layer.

where ζ is the contact potential difference between
the metal and the semiconductor. Figure 1 represents
the case when $\zeta > 0$ and we shall restrict ourselves to
this special case.

Figure 2 shows the possible energy band diagrams
for a metal covered by a thin semiconducting layer
($L \lessgtr l$) in the presence of a surface charge on its
outer surface ($\sigma_L \neq 0$). The coordinates in the diagrams are selected so that the energies are reckoned
from the Fermi level, the position of which is determined by the nature of the metal. Here ε is the
thermionic work function of the layer representing the
position of the Fermi level at the outer surface of the
layer.

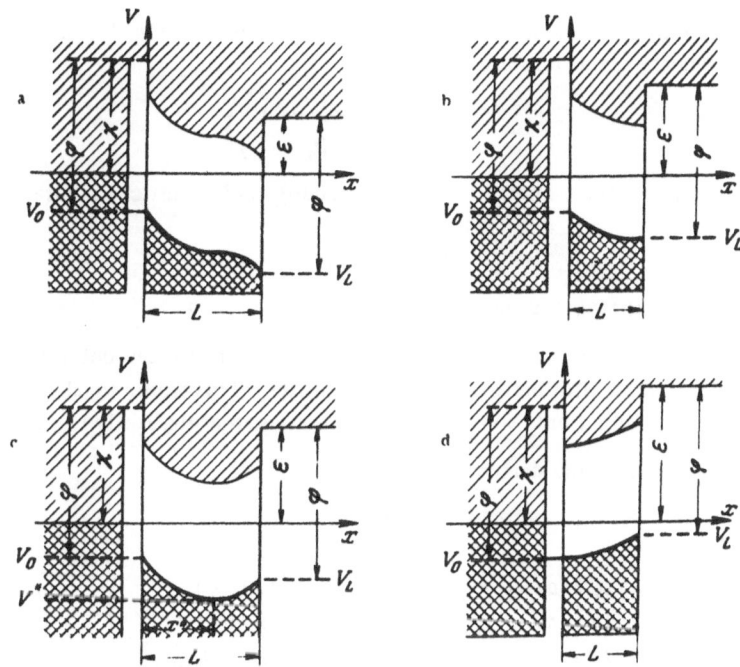

Fig. 2. Energy-band models for a metal covered with a thin semiconductor layer in the presence of a surface charge.

It can be shown that the four diagrams in Fig. 2 represent all the possible cases (when $\zeta > 0$). Figures 2a
and 2b represent a positively charged surface ($\sigma_L > 0$) and Figs. 2c and 2d represent a negatively charged surface ($\sigma_L < 0$). We shall show later that Figs. 2a and 2c represent a fairly thick film while Figs. 2b and 2d correspond to the case of a thin film. In Figs. 2a, 2b, and 2c the metal is charged negatively with respect to the
layer ($\sigma_0 < 0$, and electrons are transferred from the layer to the metal); in the case represented by Fig. 2d,
the metal is charged positively with respect to the layer ($\sigma_0 > 0$, electrons are transferred from the metal to
the layer).

At a given temperature and pressure, the specific adsorption of the layer for molecules of a given type
(i.e., the surface coverage θ with molecules of this type in equilibrium with the gas phase) and the catalytic
activity of the surface for a given reaction (i.e., the rate g of this reaction at a given temperature and given

partial pressures of gases taking part in the reaction) are determined uniquely, as shown in the electron theory of catalysis [2], by the Fermi level position ε on the surface of the layer

$$\theta = \theta (\varepsilon), \quad g = g(\varepsilon). \tag{1}$$

When $L < l$ we have

$$\varepsilon = \varepsilon (L, \chi) \tag{2}$$

and consequently

$$\theta = \theta (L, \chi), \qquad g = g(L, \chi). \tag{3}$$

The problem of determining the function (3) reduces in the final analysis to the determination of the function (2). The latter can be found by solving Poisson's equation with boundary conditions corresponding to the band model considered. We shall express the energies in units of kT and distances in units of l. Poisson's equation in dimensionless variables has the form

$$d^2V/dx^2 = \rho (V), \tag{4}$$

where $V = V(x)$ is the potential energy of the electron (the thick curve in Fig. 2), and $\rho(V)$ is the dimensionless volume charge. The boundary conditions are (cf. Fig. 2):

$$\left. \begin{array}{l} (dV / dx)_{x=L} = - \sigma_L \\ V (0) = - (\varphi - \chi) \end{array} \right\}, \tag{5}$$

where σ_L is the dimensionless surface charge. From the condition of electrical neutrality of the metal–layer system it follows that

$$(dV / dx)_{x=0} = \sigma_0, \tag{6}$$

where σ_0 is the charge on the metal in dimensionless units.

Equations (4) and (5) represent the complete system of equations for the determination of the function (2). The solution of this system meets with mathematical difficulties. Therefore, we shall now consider not the form of the functions (2) and (3) but only some of their properties in a qualitative sense.

Therefore, we shall consider the derivatives

$$\left. \begin{array}{l} (d\varepsilon / dL)_{P,T,\chi} \\ (d\varepsilon / d\chi)_{P,T,L} \end{array} \right\}; \tag{7}$$

$$\left. \begin{array}{l} (d\theta / dL)_{P,T,\chi} = (d\theta / d\varepsilon)_{P,T} (d\varepsilon / dL)_{P,T,\chi} \\ (d\theta / d\chi)_{P,T,L} = (d\theta / d\varepsilon)_{P,T} (d\varepsilon / d\chi)_{P,T,L} \end{array} \right\}; \tag{8}$$

$$\left. \begin{array}{l} (dg / dL)_{P,T,\chi} = (dg / d\varepsilon)_{P,T} (d\varepsilon / dL)_{P,T,\chi} \\ (dg / d\chi)_{P,T,L} = (dg / d\varepsilon)_{P,T} (d\varepsilon / d\chi)_{P,T,L} \end{array} \right\}. \tag{9}$$

We shall restrict ourselves to the determination of the signs of the derivatives (7), (8), (9). This will be done in Secs. 3, 5, and 6, respectively.

2. Initial Inequalities

To find derivatives (7) we shall use Poisson's equation (4).

First, we shall consider the cases when the curve $V = V(x)$ is monotonic (Figs. 2a, 2b, and 2d). In these cases, integration of Poisson's equation (4) with an allowance for Eq. (6) and the boundary conditions (5) gives

$$\left. \begin{array}{l} \left\{ 2 \int\limits_{V_L}^{V_0} \rho (V) \, dV + \sigma_L^2 \right\}^{1/2} = \mp \sigma_0 \\[4mm] \int\limits_{V_L}^{V_0} \left\{ 2 \int\limits_{V_L}^{V} \rho (V) \, dV + \sigma_L^2 \right\}^{-1/2} dV = \pm L \end{array} \right\}, \tag{10}$$

where (cf. Fig. 2):

$$V_L = V(L) = -(\varphi - \varepsilon) \Big\}$$
$$V_0 = V(0) = -(\varphi - \chi) \Big\} . \tag{11}$$

We note that here

$$\sigma_L = \sigma_L(\varepsilon), \quad \varepsilon = \varepsilon(L, \chi) \quad \varphi = \text{const}, \quad \chi = \text{const}, \tag{12}$$

and in Eq. (10) we should take

the upper sign if $\sigma_L > 0$ (Figs. 2a and 2b),

the lower sign if $\sigma_L < 0$ (Fig. 2d).

The value of ε can be found from Eq. (10).

Differentiating the left- and right-hand parts of Eq. (10) with respect to L, we obtain, allowing for Eqs. (11) and (12)

$$(d\varepsilon/dL)_{P,T,\chi} = -\sigma_L \{1 - \sigma_L [\rho(V_L) - \sigma_L(d\sigma_L/d\varepsilon)] I\}^{-1} \Big\}$$
$$(d\sigma_0/dL)_{P,T,\chi} = -\sigma_0^{-1}[\rho(V_L) - \sigma_L(d\sigma_L/d\varepsilon)](d\varepsilon/dL)_{P,T,\chi} \Big\} , \tag{13}$$

where

$$I = \int_{V_0}^{V_L} (dV/dx)^{-3} dV. \tag{13a}$$

It should be noted that in the cases represented by Figs. 2b and 2d [when the curve $V = V(x)$ has no inflection point], the integral I of Eq. (13a) can be represented in the following manner:

$$I = [\rho(V_L)\sigma_L]^{-1} + [\rho(V_0)\sigma_0]^{-1} + \int_{V_L}^{V_0} [\rho^2(V)(dV/dx)]^{-1}(d\rho/dV) dV. \tag{14}$$

Differentiating the left- and right-hand parts of the second of the equalities in Eq. (10) with respect to χ we obtain, taking into account Eqs. (11), (12), and (13),

$$(d\varepsilon/d\chi)_{P,T,L} = \sigma_0^{-1}(d\varepsilon/dL)_{P,T,\chi}. \tag{15}$$

Let us consider now the case when $V = V(x)$ has a minimum at the point $x = x^*$ (where $0 \le x^* \le L$), as shown in Fig. 2c. In this case, Poisson's equation (4) gives

$$\left\{2\int_{V^*}^{V_0} \rho(V) dV\right\}^{1/2} = -\sigma_0 \Bigg\}$$

$$\left\{2\int_{V^*}^{V_L} \rho(V) dV\right\}^{1/2} = -\sigma_L \Bigg\}$$

$$\int_{V^*}^{V_0} \left\{2\int_{V^*}^{V} \rho(V) dV\right\}^{-1/2} dV = x^* \Bigg\} , \tag{16}$$

$$\int_{V^*}^{V_L} \left\{2\int_{V^*}^{V} \rho(V) dV\right\}^{-1/2} dV = L - x^* \Bigg\}$$

where $V^* = V(x^*)$ and, as before, we have the dependences of Eqs. (11) and (12).[†] From Eq. (16) we can find ε, x^*, V^*.

_____

[†] It can be shown that the improper integrals in Eq. (16) satisfy the Cauchy condition, i.e., they converge absolutely.

Differentiating the left- and right-hand parts of the equalities in Eq. (16) with respect to L and taking into account Eqs. (11) and (12), we obtain

$$
\begin{aligned}
(d\varepsilon \,/\, dL)_{P,T,\chi} &= -\,\sigma_L \{1 - \sigma_L \,[\rho\,(V_L) - \sigma_L \,(d\sigma_L \,/\, d\varepsilon)] \, I\}^{-1} \\
(dV^* \,/\, dL)_{P,T,\chi} &= \rho^{-1}\,(V^*) \,[\rho\,(V_L) - \sigma_L \,(d\sigma_L \,/\, d\varepsilon)] \,(d\varepsilon \,/\, dL)_{P,T,\chi} \\
(d\sigma_0 \,/\, dL)_{P,T,\chi} &= -\,\sigma_0^{-1} \,[\rho\,(V_L) - \sigma_L \,(d\sigma_L \,/\, d\varepsilon)] \,(d\varepsilon \,/\, dL)_{P,T,\chi}
\end{aligned} \Bigg\} , \tag{17}
$$

where I has the form given in Eq. (14).

Differentiating the left- and right-hand parts of (16) with respect to χ we obtain, taking into account Eqs. (11), (12), and (17)

$$
\begin{aligned}
(d\varepsilon \,/\, d\chi)_{P,T,L} &= \sigma_0^{-1}\,(d\varepsilon \,/\, dL)_{P,T,\chi} \\
(dV^* \,/\, d\chi)_{P,T,L} &= \sigma_0^{-1}\,(dV^* \,/\, dL)_{P,T,\chi}
\end{aligned} \Bigg\} . \tag{18}
$$

Considering the expressions in the right-hand parts of (13) and (17) and taking into account Eqs. (15) and (18), we get for the cases shown in Figs. 2a and 2b

$$
(d\varepsilon \,/\, dL)_{P,T,\chi} < 0, \quad (d\varepsilon \,/\, d\chi)_{P,T,L} > 0; \tag{19}
$$

in the case shown by Fig. 2c, we have

$$
\begin{aligned}
(d\varepsilon \,/\, dL)_{P,T,\chi} &< 0, \quad (d\varepsilon \,/\, d\chi)_{P,T,L} > 0 \\
(dV^* \,/\, dL)_{P,T,\chi} &< 0, \quad (dV^* \,/\, d\chi)_{P,T,L} > 0
\end{aligned} \Bigg\} ; \tag{20}
$$

in the case shown in Fig. 2d, we find

$$
(d\varepsilon \,/\, dL)_{P,T,\chi} > 0 \quad (d\varepsilon \,/\, d\chi)_{P,T,L} > 0. \tag{21}
$$

It should be noted that

$$
\left.
\begin{aligned}
&\text{in the case shown in Fig. 2a, } \rho(V_L) < 0; \\
&\text{in the cases shown in Figs. 2b, 2c, and 2d, } \rho(V_L) > 0.
\end{aligned}
\right\} \tag{22}
$$

It should be noted also that

$$
(d\sigma_L \,/\, dL) = (d\sigma_L \,/\, d\varepsilon)\,(d\varepsilon \,/\, dL), \tag{23}
$$

and that always

$$
(d\sigma_L \,/\, d\varepsilon) > 0. \tag{24}
$$

From Eqs. (13) and (17) we obtain, on the basis of Eqs. (5), (6), (19), (20), (21), (22), (23), and (24), in the case shown in Fig. 2a

$$
\begin{aligned}
\sigma_L &> 0, \quad (d\sigma_L \,/\, dL)_{P,T,\chi} < 0 \\
\sigma_0 &< 0, \quad (d\sigma_0 \,/\, dL)_{P,T,\chi} > 0
\end{aligned} \Bigg\} ; \tag{25}
$$

in the case shown in Fig. 2c

$$
\begin{aligned}
\sigma_L &< 0, \quad (d\sigma_L \,/\, dL)_{P,T,\chi} < 0 \\
\sigma_0 &< 0, \quad (d\sigma_0 \,/\, dL)_{P,T,\chi} < 0
\end{aligned} \Bigg\} ; \tag{26}
$$

in the case shown in Fig. 2d

$$
\begin{aligned}
\sigma_L &< 0, \quad (d\sigma_L \,/\, dL)_{P,T,\chi} > 0 \\
\sigma_0 &> 0, \quad (d\sigma_0 \,/\, dL)_{P,T,\chi} < 0
\end{aligned} \Bigg\} . \tag{27}
$$

3. Work Function of the Layer

We shall consider the nature of the dependence of the work function of the layer ε on its thickness L and on the work function of the metal χ which forms the substrate.

From Eqs. (19), (20), and (21), we draw the following conclusions on the nature of the variation of the work function of the layer in relation to the decrease of its thickness.

1. If the layer surface is charged positively ($\sigma_L > 0$) and the film is fairly thick (cf. Fig. 2a), we have

$$\varepsilon_0 > \varepsilon > \varepsilon^*, \qquad \rho(V_L) < 0.$$

where ε^* is defined as the root of the equation

$$\sigma_L(\varepsilon^*) = 0. \tag{28}$$

Obviously, the value of ε^* is determined by the system of surface levels at the outer surface of the layer, i.e., it depends on the nature of the surface and the partial pressures of the gases in contact with the layer surface.

As the thickness of the layer decreases, the work function of the layer ε increases, as seen from Eq. (19), approaching ε_0, i.e., the Fermi level, as the surface of the layer is displaced downwards. Then V_L, as shown in Fig. 2a, remains negative but decreases in its absolute value and consequently $\rho(V_L)$ increases because in a semiconductor with a uniformly distributed impurity we always have

$$d\rho(V)/dV > 0.$$

Then, as shown by Eq. (25), σ_L remains positive but increases in magnitude, while σ_0 remains negative but decreases, i.e., electrons are transferred from the outer surface of a layer into its interior and from there into the metal.[†]

The layer thickness at which ε and $\rho(V_L)$ reach the values

$$\varepsilon = \varepsilon_0, \quad \rho(V_L) = 0,$$

will be denoted by L_{ab}. Then the inflection point on the curve $V = V(x)$ in Fig. 2a reaches the outer surface of the layer and there is a transition from the situation in Fig. 2a to that in Fig. 2b. Obviously, the critical thickness L_{ab} can be found from Eq. (10) in which it is necessary to make the substitution $\varepsilon = \varepsilon_0$ and $L = L_{ab}$.

As the layer thickness is increased further ($L < L_{ab}$, cf. Fig. 2b) we have

$$\varepsilon > \varepsilon_0 > \varepsilon^*, \quad \rho(V_L) > 0.$$

Then, as shown by Eq. (19), the work function ε continues to increase and the Fermi level continues to shift downwards (as $L \to 0$ we have $\varepsilon \to \chi$ and $\sigma_0 \to -\sigma_L$).

The variation of the work function with the layer thickness L is shown schematically in Fig. 3a for the case $\sigma_L > 0$. The several curves in Fig. 3a correspond to different values of χ. The critical layer thickness L_{ab} at which the transition from Fig. 2a to Fig. 2b occurs is determined by the intersection of the curve $\varepsilon = \varepsilon(L)$ with the straight line $\varepsilon = \varepsilon_0$.

2. If the layer surface is charged negatively ($\sigma_L < 0$), we have

$$\varepsilon_0 < \varepsilon < \varepsilon^*, \quad \rho(V_L) > 0,$$

where ε^* is determined from Eq. (28).

If the layer is fairly thick (Fig. 2c) then, as the layer thickness is decreased, the work function ε, as before, increases in accordance with Eq. (20). Consequently, the Fermi level is shifted downwards, i.e., V_L

[†] The potential shown schematically in Fig. 2a was recently calculated by Butler [3] for the case $V_0 - V_L \ll 1$. For this case, Butler found the dependence of ε on L. Our result is identical with Butler's.

increases. The behavior of the layer is different in the following two cases, governed by the nature of the metal, the previous history of the surface, the pressure and composition of the gas phase:

a) the case when $\varepsilon^* < \chi$

b) the case when $\chi < \varepsilon^*$

We shall consider each of these two cases separately.†

(a) In the case

$$\varepsilon_0 < \varepsilon < \varepsilon^* < \chi$$

the work function ε increases and approaches the value $\varepsilon = \varepsilon^*$. Then, according to Eqs. (26), (16), and (28),

$$\sigma_0 < \sigma_L < 0,$$

and σ_0, σ_L increase, with σ_L approaching $\sigma_L = 0$. From Eqs. (16) and (20), it also follows that in the case considered we have the following relationship at the minimum on the curve $V = V(x)$ in Fig. 2c

$$(L/2) < x^* < L,$$

and as ε approaches the value $\varepsilon = \varepsilon^*$, the minimum in Fig. 2c moves up and to the right, approaching the outer surface of the layer.

The layer thickness at which ε and σ_L reach the values

$$\varepsilon = \varepsilon^*, \quad \sigma_L = 0,$$

will be denoted by L_{cb}. Then the minimum in Fig. 2c reaches the outer surface of the layer ($x^* = L$, $V^* = V_L$). On further reduction of the layer thickness ($L < L_{cb}$), the quantity σ_L changes its sign and there is a transition from the situation in Fig. 2c to that in Fig. 2b, which was considered above. Of course, the critical thickness L_{cb} can be determined from the equations (16), in which it is necessary to substitute $V^* = V_L$, $L = L_{cb}$.

(b) In the case

$$\varepsilon_0 < \varepsilon < \chi < \varepsilon^*$$

the work function ε increases approaching the value $\varepsilon = \chi$. Then, as in the previous case,

$$\sigma_0 < \sigma_L < 0,$$

and σ_0, σ_L increase approaching one another. At the minimum on the curve $V = V(x)$ we have

$$(L/2) < x^* < L,$$

and as ε approaches the value $\varepsilon = \chi$, the minimum in Fig. 2c is displaced up and to the left approaching the center of the layer, in accordance with Eqs. (20) and (16). When the values

$$\varepsilon = \chi, \quad \cdot \sigma_0 = \sigma_L,$$

are reached, then we have, from Eq. (16), $x^* = L/2$, i.e., the minimum is at the center of the layer.

On further decrease of L, we have

$$\varepsilon_0 < \chi < \varepsilon < \varepsilon^*$$

and the work function ε continues to rise. Then, according to Eqs. (26) and (16),

$$\sigma_L < \sigma_0 < 0,$$

and σ_0, σ_L continue to increase. At the minimum in Fig. 2c we have

$$0 < x^* < L/2,$$

and this point continues to move upwards and to the left; this follows directly from Eqs. (20) and (16).

---

†If the surface has acceptor-type levels only, then $\varepsilon^* = \infty$, and consequently the case $\varepsilon^* < \chi$ is not realized.

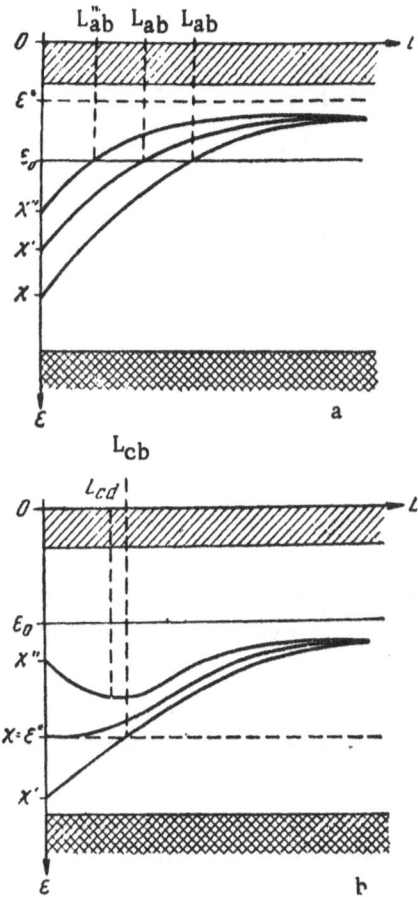

Fig. 3. Dependence of the work function on the layer thickness.

The layer thickness at which the minimum in Fig. 2c reaches the inner boundary of the layer, i.e., at which

$$x^* = 0, \quad V^* = V_0.$$

will be denoted by L_{cd}. Then, as follows from Eq. (16), $\sigma_0 = 0$, i.e., the metal is electrically neutral (the positive charge in the interior of the layer is completely compensated by the negative charge on its outer surface). When $L = L_{cd}$ there is a transition from Fig. 2c to Fig. 2d and ε reaches its maximum value $\varepsilon = \varepsilon_{max}$, where $\chi < \varepsilon_{max} < \varepsilon^*$. Obviously L_{cd} and ε_{max} can be determined from Eq. (16) by substituting $V^* = V_0$ and $L = L_{cd}$.

If the layer thickness is reduced further ($L < L_{cd}$), cf. Fig. 2d, we have

$$\varepsilon_0 < \chi < \varepsilon < \varepsilon^*,$$

and ε begins to decrease, as shown by Eq. (21), and the Fermi level is consequently displaced upwards. Then, according to Eq. (27),

$$\sigma_L < 0 < \sigma_0,$$

and σ_L decreases while σ_0 increases. Thus the metal is charged positively with respect to the layer, i.e., it loses electrons to the layer and the thinner the layer the greater the loss (as $L \to 0$, we have $\varepsilon \to \chi$ and $\sigma_0 \to -\sigma_L$).

We see that on the monotonic reduction of L the work function of the layer passes through a maximum, the physical origin of which is as follows: a fairly thick layer ($L > L_{cd}$) loses electrons from its interior to its outer surface (making it negative) and to the metal, while in the case of a thin layer ($L < L_{cd}$) the outer surface of the layer takes electrons from the interior of the layer and from the metal.

Figure 3b shows schematically the dependences $\varepsilon = \varepsilon(L)$ for various values of χ (the cases $\chi > \varepsilon^*$ and $\chi < \varepsilon^*$). The critical layer thickness L_{cb} at which there is a transition from Fig. 2c to Fig. 2b is determined by the intersection of the curve $\varepsilon = \varepsilon(L)$ and the straight line $\varepsilon = \varepsilon^*$. The critical thickness L_{cd} at which there is a transition from Fig. 2c to Fig. 2d is determined by the position of the maximum on the $\varepsilon = \varepsilon(L)$ curve.

4. Specific Adsorption of the Layer

We shall consider now the nature of the dependence of the specific adsorption θ on L and χ. We shall determine the signs of the derivatives (8).

The electron theory of chemisorption (cf. Sec. 5b in [2]) shows that

$$\left. \begin{array}{l} (d\theta/d\varepsilon)_{P,T} > 0 \text{ in the case of a donor gas} \\ (d\theta/d\varepsilon)_{P,T} < 0 \text{ in the case of an acceptor gas} \end{array} \right\}, \tag{29}$$

i.e., the specific adsorption of a semiconductor surface for a donor gas increases and for an acceptor decreases as the Fermi level position is lowered.

On the basis of Eqs. (8), (29), (19), (20), and (21), we obtain for the cases shown in Figs. 2a, 2b, and 2c

$$\left. \begin{array}{l} (d\theta/dL)_{P,T,\chi} < 0 \text{ for a donor gas} \\ (d\theta/dL)_{P,T,\chi} > 0 \text{ for an acceptor gas} \end{array} \right\} . \tag{30}$$

In the case shown in Fig. 2d, we have

$$\left. \begin{array}{l} (d\theta/dL)_{P,T,\chi} > 0 \text{ for a donor gas} \\ (d\theta/dL)_{P,T,\chi} < 0 \text{ for an acceptor gas} \end{array} \right\} . \tag{31}$$

Moreover, in all cases, we have

$$\left. \begin{array}{l} (d\theta/d\chi)_{P,T,L} > 0 \text{ for a donor gas} \\ (d\theta/d\chi)_{P,T,L} < 0 \text{ for an acceptor gas} \end{array} \right\} . \tag{32}$$

From Eqs. (30), (31), and (32), we draw the following conclusions.

1. Let us assume that the outer surface of the layer has a sufficiently high density of surface states of non-adsorptive origin. In this case, the adsorption of gas alters somewhat the absolute magnitude of the surface charge without changing its sign. In other words, the sign of σ_L is governed not by the nature of the gas being adsorbed but by the previous history of the adsorbing surface.

In this case, if the surface of a fairly thick layer is charged positively ($\sigma_L > 0$) or if it is charged negatively ($\sigma_L < 0$) but the condition $\chi > \varepsilon^*$ is satisfied, its specific adsorption for a donor gas rises monotonically and for an acceptor gas falls monotonically as the layer thickness is decreased.

If the surface of a fairly thick layer is charged negatively ($\sigma_L < 0$) and also $\chi < \varepsilon^*$, then as the thickness is reduced the specific adsorption for a donor gas passes through a maximum and that for an acceptor gas passes through a minimum.

2. Let us consider now the opposite limiting case: the outer surface of the layer has no nonadsorptive surface states. In this case, the sign of the surface charge is determined completely by the nature of the gas being adsorbed.

On adsorption of a donor gas ($\sigma_L > 0$), the specific adsorption of the layer increases monotonically as the layer thickness decreases, but on adsorption of an acceptor gas ($\sigma_L < 0$) the specific adsorption decreases, passes through a minimum and then increases again.

3. For a given nature of the layer and a given thickness L (where $L < l$), the specific adsorption of the layer for a donor gas increases and for an acceptor gas decreases with increase of the work function χ of the metal under the layer, irrespective of the sign of the surface charge σ_L on the outer surface of the layer.

5. Catalytic Activity of the Layer

Let us assume a catalytic reaction is proceeding on the outer surface of the layer. We shall determine the nature of the dependence of the reaction rate g on the layer thickness L and on the work function χ of the metal under the layer. We shall determine the signs of the derivatives (9).

According to the electron theory of catalysis, all reactions on the surface of a semiconductor can be divided into two classes: donor and acceptor reactions (cf. Sec. 6c in [2]). A reaction is classified as donor if a lowering of the Fermi level (other conditions being equal) increases the reaction rate. A reaction is classified as acceptor if a lowering of the Fermi level retards the reaction. Thus, we have by definition

$$\left. \begin{array}{l} (dg/d\varepsilon)_{P,T} > 0 \text{ for a donor reaction} \\ (dg/d\varepsilon)_{P,T} < 0 \text{ for an acceptor reaction} \end{array} \right\} , \tag{33}$$

where the symbol P stands for the sum total of the partial pressures of all the gases participating in the reaction.

From Eqs. (9), (33), (19), (20), and (21) we obtain for the cases shown in Figs. 2a, 2b, and 2c

$$\left. \begin{array}{l} (dg/dL)_{P,T,\chi} < 0 \text{ for a donor reaction} \\ (dg/dL)_{P,T,\chi} > 0 \text{ for an acceptor reaction} \end{array} \right\} . \tag{34}$$

In the case shown in Fig. 2d we have:

$$\begin{aligned} (dg/dL)_{P,T,\chi} &> 0 \text{ for a donor reaction} \\ (dg/dL)_{P,T,\chi} &< 0 \text{ for an acceptor reaction} \end{aligned} \Bigg\} . \qquad (35)$$

Moreover, in all cases we have:

$$\begin{aligned} (dg/d\chi)_{P,T,L} &> 0 \text{ for a donor reaction} \\ (dg/d\chi)_{P,T,L} &< 0 \text{ for an acceptor reaction} \end{aligned} \Bigg\} . \qquad (36)$$

From Eqs. (34), (35), and (36), we draw the following conclusions.

1. If the value of L is sufficiently large and the layer surface is charged positively ($\sigma_L > 0$) or if it is charged negatively ($\sigma_L < 0$) but the condition $\chi > \varepsilon^*$ is satisfied, then the catalytic activity of the layer in a donor reaction increases monotonically and in an acceptor reaction decreases monotonically on the reduction of L.

If the value of L is sufficiently large and the layer surface is charged negative ($\sigma_L < 0$) and also $\chi < \varepsilon^*$, then the catalytic activity of the layer in a donor reaction passes through a maximum and in an acceptor reaction passes through a minimum when the value of L is decreased monotonically.

It should be noted that the surface charge σ_L is determined by the content of all the molecules present on the surface and taking part in the reaction or being products of the reaction, as well as by the surface defects due to the previous history of the sample.

2. Let us assume that a reaction has two consecutive stages: one acceptor, the next donor. As the layer thickness is varied the rate-limiting role may be transferred from one stage to the other.

Thus, for example, if the reaction rate on a layer of given thickness is limited by the donor stage, then on a thinner layer (all other conditions being equal) the acceptor stage of the reaction may become the limiting one. This may affect the sequence of the reaction.

3. Let us assume that a reaction proceeds in two parallel directions, one of which is donor (rate g_D) and the other acceptor (rate g_A). The ratio g_D/g_A characterizes the selectivity of the catalyst. Obviously, the selectivity of the layer varies with its thickness.

Thus, for example, if on a layer of a given thickness the reaction proceeds mainly in the acceptor direction ($g_A \gg g_D$) then the reduction of the layer thickness may alter the direction to the donor ($g_D \gg g_A$), provided all other conditions are equal. Thus variation of the layer thickness may alter even the direction of the reaction.

4. For a given nature of the layer and a given thickness L (where $L < l$), the catalytic activity of the layer in a donor reaction increases and in an acceptor reaction decreases with increase of the work function of the metal χ underlying the layer.

6. Conclusions

In most cases met with in practice, the semiconducting sheath covering a metal is produced by oxidation of the metal. It represents an oxide layer, the thickness of which can frequently be controlled. An example of this is cuprous oxide grown on a copper matrix.

By varying the thickness of the sheath, as shown above, we can to a certain extent control the specific adsorption, the catalytic activity, and the selectivity of the sample. It would be interesting to check experimentally this theoretical prediction. It is important that in such a check the sheath thickness L should be smaller than the screening length, i.e., $L < 10^{-4} - 10^{-5}$ cm, but the sheath should not be too thin ($L > 10^{-6}$ cm) in order to be considered as an independent phase. It is also important that when the layer thickness is varied, all the other conditions which could distort or mask the effect investigated should be kept constant.

LITERATURE CITED

1. S. Z. Roginskii, Problemy Kinetiki i Kataliza **4**, 187 (1940).
2. F. F. Vol'kenshtein, Electron Theory of Catalysis on Semiconductors (Fizmatgiz, 1960).
3. H. N. Butler, J. Chem. Phys. **35**, 636 (1961).

EFFECT OF LOW-MELTING GLASSES OF THE As−S−I SYSTEM ON THE CURRENT−VOLTAGE CHARACTERISTICS OF SILICON p−n JUNCTIONS

V. I. Gaman, A. A. Sirotkin, and
V. M. Stenina

V. V. Kuibyshev State University, Tomsk

One of the most important problems in the present-day technology of semiconducting devices is the reduction of their size. For this purpose, it is necessary to develop methods giving reliable protection from the surrounding medium and ways of stabilizing the parameters of devices which are not encased. One method of protecting a surface is to cover the device with glass of the As−S−I system. Investigations carried out by Flaschen, Pearson, and Kalnins [1] on diffused silicon p−n junctions proved the method to be successful. After coating with glass the reverse currents of p−n junctions with well-treated surfaces decreased by one or two orders of magnitude under a voltage of 80% of the breakdown value. When a device with a dirty surface was coated with glass, the reverse currents increased and the breakdown voltage decreased. However, when a dirty device was held in molten glass at 275°C for several tens of hours, the reverse currents and the breakdown voltage assumed the same values as for clean samples.

We carried out a similar study on alloyed p−n junctions. The devices were coated with a glass of the following composition (in wt.%): 24 As, 67 S, 9 I. The glass was boiled in a nonoxidizing medium at 500-600°C. The initial components were of high purity. Measurements showed that the loss-angle tangent of the glass was small and decreased as the frequency was increased from 30 to 10^4 cps (Fig. 1). The permittivity of the glass was 6.5.

The glass coating was applied to p−n junctions produced by fusing aluminum into n-type silicon of 10-15 $\Omega \cdot$ cm resistivity. A fused-in aluminum lug served as one of the current leads. The other lead was spark-welded to the gold support of the device by means of special apparatus [2]. After the attachment of the lead, the samples were etched in a mixture of nitric and hydrofluoric acids and dried for 2-3 hours at 150°C.

The reverse static current—voltage characteristics of the devices were recorded in darkness. Then the devices were immersed in molten glass at 250-300°C and held there for 1 min. The current—voltage characteristics were recorded after coating and the samples then kept in a thermostat at 130-150°C for 30-50 hours. Next, the samples were subjected to conditions of tropical humidity for 72 hours and then to thermal shock. Thermal cycling was applied in the temperature range from -60 to +130°C: the samples were kept for 30 min at -60°C, 10 min at room temperature, and 30 min at +130°C, and the cycle was repeated twice.

In most of the samples, the reverse currents increased after coating with glass (Fig. 2). After heat treatment for 30-50 hours at 130-150°C, the reverse currents returned to their original values or even fell below

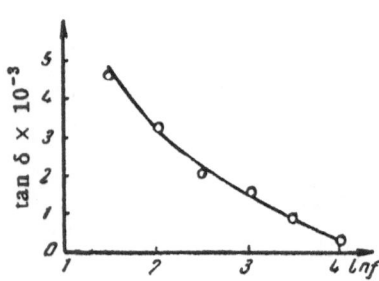

Fig. 1. Frequency dependence of tan δ for a glass of the As−S−I system.

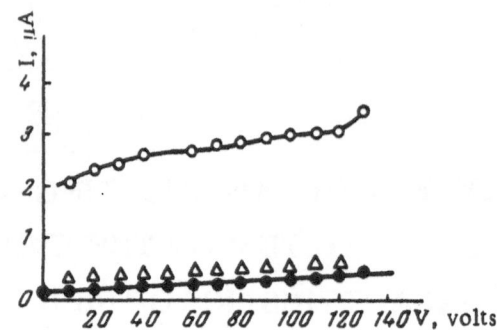

Fig. 2. Reverse current−voltage characteristics of a silicon diode (sample No. 88): ● - before coating with glass; ○- after coating with glass; △ - after drying for 30 hours.

these values (Fig. 2). In the case of samples with well-treated surfaces, after drying the reverse currents usually returned to the original value or fell by a factor of 1-2. For samples with dirty surfaces (currents of the order of microamperes), the reverse currents, measured after coating with glass and drying, decreased by 1-2 orders of magnitude, i.e., they assumed the same values as for clean samples (Fig. 3). The breakdown strength also increased and the reverse current drift, observed in some samples, disappeared.

After testing under tropical humidity conditions, the reverse characteristics were not altered markedly or were slightly improved (Fig. 4). This indicates that the glass coating is waterproof.

Only 50% of the devices withstood a thermal shock, the reverse characteristics of the remaining unaltered. The reverse characteristics of the 50% that failed deteriorated very markedly although no changes in the glass coating could be seen. The places most vulnerable to thermal-shock tests were the junctions of the glass with the metal leads, largely because (1) the technique of coating semiconductors with glass has not yet been perfected and (2) the extent to which the leads were sealed in varied from sample to sample. It is probable that microcracks were formed at the junctions of the glass and aluminum through which moisture entered during thermal cycling. To reduce the rate of failure in thermal shock tests, it is necessary, firstly, to improve the technique of coating with glass, and, secondly, to test the suitability of other metals, for example gold, as lead materials.

The influence of glasses of the As−S−I system on the current−voltage characteristics of silicon p−n junctions can be understood in terms of excess surface currents which govern the reverse current. The reverse current of a silicon p−n junction can be represented as a sum of three components:

$$I = I_d + I_r + I_n,$$

where I_d is the diffusion current; I_r is the generation-recombination current; I_n is the surface leakage current. The surface leakage current is the largest component of the reverse current when the current etching methods are employed [3]. The diffusion and generation-recombination components of the reverse current may be observed only under certain conditions [3]. The excess surface currents of most silicon devices may be due to the presence of an "anomalous" surface channel [4]. The nature of the "anomalous" channel is still not clear but it has been established that it appears in the case of a high density of surface states having positive or negative charge. In the case of a sufficiently large negative charge on silicon the "anomalous" channel transforms into a normal channel due to an inversion layer. Under conditions of high humidity, the surface leakage current may be due to "quasi-ionic" conductivity [5].

In most of our samples, the reverse currents increased proportionally to the applied voltage. Such a current−voltage dependence is characteristic of p−n junctions with the "anomalous" channel [4]. The surface leakage current due to the "anomalous" channel varies with the magnitude of the corresponding surface charge.

148

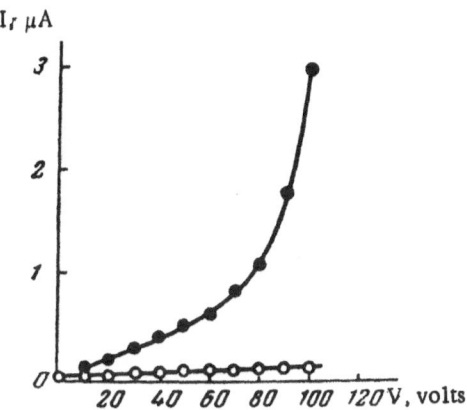

Fig. 3. Reverse current—voltage characteristics of a silicon diode (sample No. 91): ● - before coating with glass; ○ - after coating with glass and drying for 25 hours.

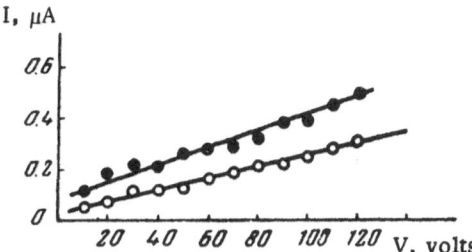

Fig. 4. Reverse current—voltage characteristics of a silicon diode (sample No. 88): ●- before testing under tropical humidity conditions; ○ - after tropical humidity and thermal shock testing.

The surface charge is governed by the chemisorbed molecules which have "strong" acceptor or donor bonds with the surface [6]. Apart from the molecules strongly bound to the surface, there are always molecules in a state of "weak" chemisorption and physical adsorption. Molecules of the latter type may become chemisorbed on being supplied with a definite activation energy. When the devices are immersed in molten glass, it is probable that the relationships between the various types of adsorption are altered and therefore the surface charge is affected. This may explain the change in the reverse current immediately after coating with glass. On the other hand, the glass acts as a getter.

Prolonged exposure to elevated temperatures gives rise to the absorption by the glass of the adsorbed molecules, and reduction of both the surface charge and the reverse current. If the surface of the device is sufficiently clean, then there should be no increase of the reverse current immediately after coating with glass or this increase should be very small. This was confirmed experimentally.

Apart from the components of the surface current just enumerated, at high reverse bias there may be another component due to impact ionization on the surface of the p—n junction. This ionization results in a pronounced rise of the reverse current with increase of the applied voltage, and leads to surface breakdown. The surface breakdown voltage is governed by the width of the depleted region of the p—n junction at the surface, which in turn depends on the magnitude and sign of the surface charge [7]. Coating with glass alters considerably the voltage at which a sharp rise of the reverse current begins (Fig. 3). This indicates a change of the surface charge. The chemisorbed molecules are removed from the surface and the width of the depleted region of the p—n junction on the surface approaches the width in the interior. This increases the surface breakdown voltage right up to the value representing volume breakdown. In our devices the volume breakdown voltage was in the range 120-180 V.

From our experiments, we conclude that glasses of the As—S—I system act as getters which absorb surface contamination. Glasses of this system are waterproof and have low dielectric losses, properties which recommend these glasses as coatings to protect semiconductor devices from their environment.

Concluding the authors express their gratitude to B. V. Makarkin for measuring the dielectric properties of the glasses.

LITERATURE CITED

1. S. S. Flaschen, A. D. Pearson, and I. L. Kalnins, J. Appl. Phys. 31, 2 (1960).
2. N. B. Brandt, Pribory i Tekh. Eksp. No. 2, 138 (1956).
3. D. J. Sandiford, J. Appl. Phys. 30, 1981 (1959).
4. R. Solomon, J. Appl. Phys. 31, 1791 (1960).

5. O. Jäntsch, Z. Naturforsch. 15a, 141 (1960).
6. B. F. Vol'kenshtein, Electron Theory of Catalysis on Semiconductors (Fizmatgiz, 1960).
7. C. G. B. Garrett and W. H. Brattain, J. Appl. Phys. 27, 299 (1956).

EFFECT OF THE ADSORPTION OF CERTAIN AMINES ON A SEMICONDUCTOR SURFACE ON THE PRINCIPAL PARAMETERS OF GERMANIUM TRANSISTOR TRIODES

G. A. Kataev, V. A. Presnov,
E. N. Batueva, Yu. G. Kataev, and
L. A. Lyuze

V. V. Kuibyshev State University, Tomsk

The physico-chemical state of a germanium surface in relation to its electrical properties is of considerable interest both in practice and in theory.

This state is governed by complex chemical and adsorption compounds, present on the surface, which determine the density and positions of the energy levels of impurity centers, as well as the total surface charge on which, in turn, depend the conditions of recombination and the electrical conductivity on the semiconductor surface.

There is little published information on how the changes in the composition and structure of the surface compounds affect the changes of the surface charge and to what extent the particles present on the surface may act as impurity centers [1-4].

This information is concerned mainly with the influence of the processes of oxidation, hydration, and dehydration of the surface oxides [3] on the parameters of semiconducting devices.

We investigated the effect of the adsorption of certain amines of the aliphatic and aromatic series on the surface of germanium transistors on the amplification factor α and the reverse collector currents I_{co}.

The experimental results indicate that the adsorbed amines considerably alter the principal parameters of germanium devices.

1. Experimental Technique

The tests were made on P-5 transistors. Unprotected transistors were subjected to weak etching in hydrogen peroxide (3 min), washing in deionized water in a quartz vessel, and drying in circulating air for 2 hours at a controlled temperature of 90°C. The triodes treated in this way were cooled for 30 min in a desiccator containing calcium oxide.

After measuring the transistor parameters (α, I_{co}, and leakage currents) using well-known circuits [5], the transistors were treated with amines. In the case of liquid amines with high boiling points, the devices were immersed in a hot (85-90°C) amine (aniline, dimethylaniline, aniline black, quinoline, triethylamine) for 1 hour. In the case of amines solid under normal conditions, we used their alcohol solutions and again the triodes were immersed in the solution for 1 hour at the boiling point of the solvent (ethyl alcohol). Treatment with ethyl alcohol alone did not produce any noticeable or stable changes of the parameters and therefore it acted here as an inert solvent.

After treatment with amines, the transistors were dried at a controlled temperature of 90°C for 2 hours and cooled in a desiccator. Then their parameters were measured. The same treatment, except for immersion in an amine or an amine solution, was applied to the control samples.

All the reagents used were of chemically pure or analytical grade; some amines were subjected to additional purification by distillation. Each treatment was applied to a batch of 3-6 transistors. Table 1 lists the average values of α and I_{co}. The reproducibility of the measured parameters can be seen in Table 2.

TABLE 1

Amine	Dissociation constant	I_{co}, μA		Change, %	α		Change, %
		Before adsorption	After adsorption		Before adsorption	After adsorption	
Hexamethylenediamine	—	42.2	10.9	-76.1	0.965	0.936	-3.0
Triethylamine	$5.65 \cdot 10^{-4}$	31.0	15.1	-51.0	0.954	0.960	+0.42
Ammonia	$1.79 \cdot 10^{-5}$	20.0	45.0	+109.0	0.941	0.989	+6.5
p-Phenylenediamine	$\begin{cases} 1.1 \cdot 10^{-8} \\ 3.5 \cdot 10^{-12} \end{cases}$	20.5	18.2	-10.7	0.957	0.985	+2.9
p-Toluidine	$1.18 \cdot 10^{-9}$	25.0	8.0	-67.5	0.950	0.940	-1.5
Dimethylaniline	$1.15 \cdot 10^{-9}$	29.0	12.7	-56.0	0.830	0.723	-13.0
Benzidine	$\begin{cases} 9.3 \cdot 10^{-10} \\ 5.6 \cdot 10^{-11} \end{cases}$	6.5	7.0	+0.7	0.955	0.945	-1.0
Aniline	$3.82 \cdot 10^{-10}$	14.5	11.0	-25.0	0.885	0.842	-4.8
β-Naphthylamine	$1.29 \cdot 10^{-13}$	34.0	10.7	-70.0	0.800	0.800	0.0
Diphenylamine	$7.6 \cdot 10^{-14}$	72.5	14.5	-79.5	0.930	0.897	-3.8
Aniline black	—	17.0	5.2	-74.0	0.851	0.964	+13.2

Note: + denotes increase, - denotes decrease.

TABLE 2

I_{co}, μA		Change, %	α		Change, %
Before adsorption	After adsorption		Before adsorption	After adsorption	
Dimethylaniline					
28.0	16.0	-43.0	0.800	0.690	-13.5
32.0	10.0	-68.0	0.780	0.650	-16.5
27.0	12.0	-56.0	0.910	0.830	- 9.0
Triethylamine					
29.0	13.0	-59.0	0.940	0.950	+11.0
25.0	12.5	-50.0	0.935	0.930	- 0.5
29.0	15.0	-49.0	0.980	0.985	+0.51
41.0	20.0	-51.0	0.960	0.975	+1.50

The results plotted in the various figures represent measurements of the parameters of one transistor which showed a typical dependence of the parameters on time for treatment with a given amine.

2. Experimental Results and Discussion

The experimental results of our study of the influence of amine adsorption on the principal parameters of germanium transistors may be summarized as follows (cf. Table 1, Figs. 1 and 2).

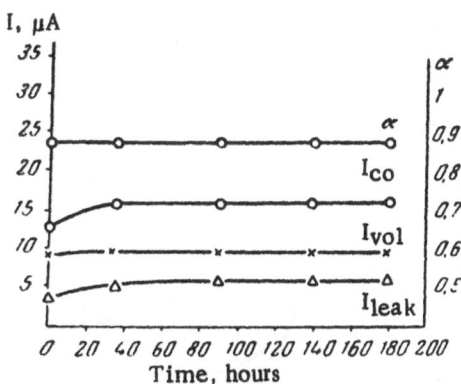

Fig. 1. Influence of treatment in triethylamine on the triode parameters.

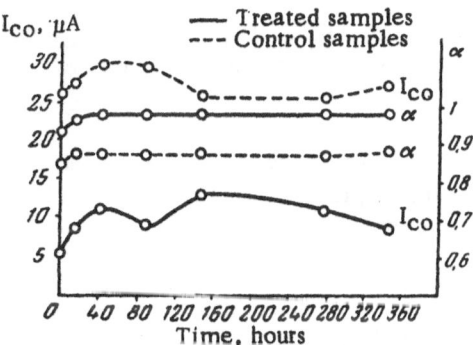

Fig. 2. Influence of treatment in benzidine on the triode parameters.

Firstly, the adsorption of amines always reduces the reverse collector currents. This reduction is easily reproduced and its magnitude depends on the specific properties of the amines.

Secondly, the adsorption of amines alters the amplification factor, which may increase or decrease depending on the alkalinity of the amines.

Thirdly, the strength of the adsorption bond, which may be represented by the stability of the transistor parameters with time under normal conditions, is determined by the nature of the amine (Figs. 1 and 2).

Finally, the treatment of a germanium surface with amines by the technique described above alters the surface charge — always making it less negative [6].

These results may be explained by taking into account the facts that, firstly, the adsorption of amines alters the surface charge of a semiconductor and this affects the device parameters, and secondly, this adsorption reduces the leakage currents by forming donor—acceptor bonds with those ions which are capable of migration on the surface in an electric field (p—n junction).

The adsorption of amines on the semiconductor surface is the result of donor—acceptor interactions.

In all amines, the nitrogen atom has an unshared pair of electrons capable of forming bonds with substances exhibiting electron-acceptor properties.

The first important consequence of the donor—acceptor interactions is the change in the surface charge. This is apparently due to two causes. Firstly, the amine forms bonds with the hydrated germanium surface giving rise to an external double electric field which compensates the negative surface charge. This interaction may be accompanied by the formation of a weak hydrogen bond:

$$-\overset{|}{\underset{|}{Ge}}-\overset{-\delta}{O}\ldots\overset{+\delta}{H}\ldots\overset{\overset{R}{|}}{\underset{\overset{|}{R}}{N}}-R \qquad (1)$$

or by an acid—base interaction with a complete transfer of a proton to the nitrogen atom:

$$-\overset{|}{\underset{|}{Ge}}-\bar{O}\left[H:\overset{\overset{R}{|}}{\underset{\overset{|}{R}}{N}}-R\right]^{+}. \qquad (2)$$

Moreover, an electron of the unshared nitrogen pair may be transferred directly to electron—acceptor centers on the surface (for example, holes) forming radical-cations and this again partly compensates the negative surface charge:

$$R - \underset{\underset{R}{|}}{\overset{\overset{R}{|}}{N}} : + \overrightarrow{L} \rightarrow \left[R - \underset{\underset{R}{|}}{\overset{\overset{R}{|}}{N}} . \right]^{+} \dots L\bar{e},$$

where L represents the lattice.

The change in the surface charge alters the φ_s potential and consequently the transistor parameters.

The changes in the amplification factor and the reverse collector currents due to the change in the surface potential φ_s are represented, respectively, by curves 2 and 1 in Fig. 3 [7].

Figure 3 shows that when the negative surface charge of an n-type semiconductor is altered, the amplification factor varies in accordance with curve 2 passing through a minimum approximately at $\varphi_s = 0$ (which corresponds to the maximum of the surface recombination velocity in the curve of Stevenson and Keys [8]).

The dependence of I_{co} on the magnitude of the surface charge is given by curve 1 in Fig. 3.

The latter curve indicates that as the negative surface charge is reduced, the reverse current through the p—n junction, due to minority carriers, decreases because the reduction of the negative charge lowers the density of holes in the surface layers of the semiconductor.

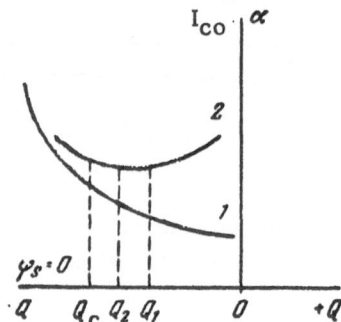

Fig. 3. Dependence of I_{co} and α on the surface charge.

These explanations are confirmed by the experimental results: the amines with more strongly marked alkaline properties (stronger donor functions of the nitrogen atom, and dissociation constants $\simeq 10^{-4} - 10^{-8}$) increase the amplification factor as a result of a large change of the surface charge Q; the amines which are weakly alkaline (dissociation constants $10^{-8} - 10^{-9}$) reduce the amplification factor. This can be understood in terms of the small change in the surface charge on adsorption: for example, the change of the charge from Q_C to Q_2 or Q_1 (Fig. 3).

The second consequence of the formation of the donor—acceptor bonds is the locking of hydrogen ions belonging to hydrated germanium oxides (or other impurity ions), which reduces considerably the leakage currents that are due to migration of ions in the electric field of the p—n junction. This is confirmed by the fact that the reverse currents are reduced after treatment with amines and a large part of the reduction of these currents is due to smaller leakage currents.

Among the simplest amines, only ammonia increases I_{co} due to the increase of the leakage currents. This is because ammonia is easily soluble in water and its adsorption on the germanium surface is therefore accompanied by the adsorption of water vapor from air, which produces NH_4OH on the transistor surface. The latter compound dissociates

$$NH_4OH \rightleftarrows NH^+_4 + OH^-$$

giving rise to large numbers of OH^- and NH^+_4 ions which take part in charge transport (shunting of the p—n junction).

This behavior is impossible in the case of organic amines because of their hydrophobic nature and the large dimensions of their cation-radicals which considerably reduce their mobility (particularly in the case of amines of the aromatic series).

Finally, it should be noted that the transistors treated with amines were more stable than the control samples (cf. Figs. 1 and 2). Particularly good stabilization of the devices with satisfactory values of the parameters was obtained by treating the triodes with aniline black, triethylamine and hexamethylenediamine.

This indicates that the adsorption bonds of these three amines are quite strong. The adsorption of the other amines (β-naphthylamine, diphenylamine, benzidine, and others) is more reversible since even small variations of temperature and humidity alter the transistor parameters, probably due to partial desorption.

3. Conclusions

1. The influence of the adsorption of aliphatic and aromatic amines on the principal parameters of germanium transistors was investigated.

2. It was found that the change in the transistor parameters on adsorption of amines is related to the change in the surface potential, on the one hand, and to the change of the magnitude of the leakage currents, on the other.

3. The observed behavior can be explained by taking into account the donor—acceptor interactions of the adsorbed molecules with the germanium surface.

LITERATURE CITED

1. W. H. Brattain and J. Bardeen, Bell System Tech. J. 32, 1 (1953).
2. A. R. Hutson, Phys. Rev. 102, 381 (1956).
3. G. A. Barnes and P. C. Banbury, in the collection: Physics of Semiconductor Surfaces, ed. by Pikus [Russian Translation] (IL, 1959), p. 64-5;
   S. G. Ellis, ibid., p. 103-124;
   J. Wallmark and P. Johnson, ibid., p. 342-358;
   J. Wallmark, ibid., p. 320-341.
4. J. A. Dillon and H. E. Farnsworth, Phys. Rev. 99, 1643 (1955).
5. A. Ya. Fedotov and Yu. V. Shmartsev, Transistors (Izd. "Sov. radio," 1960).
6. V. A. Presnov and L. L. Lyuze, this volume, p. 156.
7. V. A. Presnov and G. A. Kataev, Semiconducting Devices and their Applications, ed. by A. Ya. Fedotov (1962), No. 8, p. 26.
8. A. Stevenson and R. Keys, Physica 20, 1041 (1954).

EFFECT OF AMINE ADSORPTION ON THE
SURFACE CHARGE OF GERMANIUM

V. A. Presnov and L. L. Lyuze

V. V. Kuibyshev State University, Tomsk

The surface properties of semiconductors exert a considerable influence on the parameters of the semi-conducting device made from them. One factor which alters surface properties is the surface charge. After etching, a germanium surface usually has a negative charge [1]; therefore, it is desirable to find substances which reduce this charge. Consequently, we investigated a series of amines, ranging from those with strong alkaline properties to others which were weakly acidic.

Measurements were made of the surface potential of high-resistivity germanium before and after treatment with the amines. Since the surface potential is governed by the charge at the "slow" surface states, a knowledge of the magnitude and sign of this potential can give us information on the surface charge.

1. Experimental Technique

To measure the surface potential, we used the ac field effect. The block diagram of the apparatus [2] is shown in Fig. 1. We avoided gluing the sample down since this makes it difficult to replace the sample and is liable to contaminate its surface. Instead, a special holder was developed (Fig. 2). The holder should, first of all, ensure rigid mounting of the sample; and secondly, it should permit its rapid replacement. The rigid mounting of the sample is essential because otherwise it vibrates under the action of the alternating current, thereby considerably distorting the signal.

A mica leaf, 20-30 μ thick, was placed on a polished brass plate (1 in Fig. 2). To avoid charge leakage across the surface, paraffin was poured into the space between the holder base and the mica (2 in Fig. 2). The sample was fixed below the mica leaf. Samples were cut from a single crystal of n-type germanium ($\rho = 31$ $\Omega \cdot cm, \tau = 200 \mu sec$); the sample dimensions were: 12-18 mm long, 2.5-4 mm wide and 0.2-0.3 mm thick. The contacts were soldered with tin. The measurements were made in air, and the frequency of the applied field was 60-140 cps.

The modulation of the conductivity by the applied electric field was displayed as a trace on the oscillograph screen, which in fact represented the dependence of the change in the surface conductivity on the surface charge. The traces were photographed and then their ordinates were measured from their minimum to the point representing zero field at the electrode. This point was in the middle of the abscissa segment defined by the extreme points of the trace. From the ordinate, we calculated the change in the potential drop across the sample ΔV. Hence, we found the change in the conductivity per unit area ΔG from the formula [3]

$$\Delta G = \frac{\Delta V L}{I \cdot R^2 \cdot w} ,$$

(1)

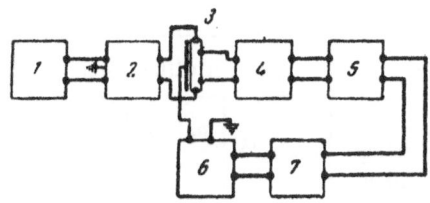

Fig. 1. Block diagram of the apparatus used in measurements of the field effect. 1) DC source; 2) filter; 3) sample; 4) differential amplifier; 5) cathode-ray oscillograph; 6) modulator; 7) phase shifter.

Fig. 2. Sample holder. 1) Brass plate; 2) paraffin; 3) germanium sample; 4) mica.

where ΔV is the change in the potential drop; L is the sample length; I is the current flowing through the sample; R is the resistance and w is the width of the sample.

From the values of ΔG, we determined the surface potential V_s using a curve showing the dependence of ΔG on V_s. This curve was obtained by calculation using previously published tables [5]. Since the observed band curvature was relatively small, the mobilities used in the calculation of ΔG were assumed to be equal to the volume values [4].

As two values of the surface potential correspond to every value of ΔG, it was necessary in each case to determine with which branch of the $\Delta G(V_s)$ curve we are dealing. It was difficult to determine this directly since the phase shift is altered during tuning. The minimum of the trace may appear either with a positive voltage at the field electrode or with a negative one, depending whether there is an inversion layer or an enriched layer at the surface.

According to Bardeen et al. [6, 7], we can determine the sign of the branch from the effective mobility. For samples with relatively high mobility due mainly to one type of carrier, it is found that the effective electron mobilities are of the order of $(1-1.5) \times 10^3$ $cm^2 \cdot V^{-1} \cdot sec^{-1}$ and the hole mobilities of the order of $(0.3-0.6) \times 10^3 cm^2 \cdot V^{-1} \cdot sec^{-1}$. The sign of the effect can be determined also from the sign of the change in the surface conductivity on the application of a constant field. This change can be measured with a galvanometer.

In the case of samples for which the oscillograms of the field effect were almost symmetrical, i.e., those whose surface potential had values corresponding to the conductivity minimum, the conductivity rose on the application of both a positive and a negative field to the field electrode.

In the case of samples for which the oscillograms had a branch in addition to a minimum, a positive field raised the conductivity if the branch represented electronic conductivity. However, if the excess conductivity was mainly due to holes, a positive field reduced the conductivity. Thus, it was possible to determine unambiguously the sign of the surface potential.

Before the measurements, the samples were etched in boiling H_2O_2 and washed in deionized water; then they were dried in a desiccator. Next, the surface potential was measured and the samples were treated with an amine. Then, after drying, the surface potential was measured again.

2. Experimental Results

The results of the measurements and calculations are given in the table, from which it is clear that the sample surface, after etching and drying, was negatively charged ($V_s < 0$, upward band curvature for n-type samples). This charge appears directly as a result of the oxidation of germanium with oxygen which, having a strong affinity for electrons, charges the surface negatively.

The magnitude of the initial band curvature varies within the limits 3 ± 1 kT/e. The scatter of the band curvature values is due to the instability of the freshly etched surface during the measurements and daily variations of the humidity of air. After treatment, there was still some scatter of the surface potential values since it was very difficult to obtain identical conditions on surfaces of different samples.

Effect of Surface Treatment of Ge on Surface Potential

	Triethylamine				Hexamethylenediamine				Aniline black		
Sample No.	Before treat- ment	After treat- ment	Change of V_s	Sample No.	Before treat- ment	After treat- ment	Change of V_s	Sample No.	Before treat- ment	After treat- ment	Change of V_s
1	-2.9	+0.8	3.7	5	-3.1	-0.2	2.9	8	-3.1	-1.3	1.8
2	-5.2	+1.05	6.25	6	-2.3	+0.8	3.1	9	-3.2	-1.3	1.9
3	-5.4	+1.15	6.55	7	-3.3	-0.2	3.2				
4	-6.1	+2.1	8.2								

Effect of Surface Treatment of Ge on Surface Potential

	Diphenylamine				β-Naphthylamine	
Sample No.	Before treat- ment	After treat- ment	Change of V_s	Sample No.	Before treat- ment	After treat- ment
10	-2.3	-3.1	0.8	14	4.4	-3.7
11	-3.1	-3.7	0.6	15	4.6	-4.1
12	-3.0	-3.4	0.4	16	4.1	4.6
13		-3.8		17	4.2	4.3
				18	4.3	4.2

On the adsorption of the amines with strongly alkaline properties (triethylamine, hexamethylenediamine, aniline black) a considerable reduction of the negative charge was observed. The stronger the alkalinity of an amine, the greater the reduction of the negative surface charge. The greatest effect was produced by triethylamine (V_s changed by 6 kT/e), whereby the charge changed from a large negative to a small positive value. A somewhat smaller reduction of the charge was produced by hexamethylenediamine (V_s changes by 3 kT/e) and aniline black (V_s changes by 2 kT/e).

Diphenylamine increased the negative charge somewhat but the charge returned to its initial value during measurements. This was even more strongly marked in the case of samples treated with β-naphthylamine. These two amines have very weak alkaline properties; therefore they are weakly bound to the germanium surface and are easily desorbed.

LITERATURE CITED

1. J. Wallmark, RCA Rev. 18, 255 (1957).
2. O. V. Sorokin, Pribory i Tekh. Eksp. No. 2, 68 (1959).
3. H. C. Montgomery and W. L. Brown, Phys. Rev. 103, 865 (1956).
4. J. R. Schrieffer, Phys. Rev. 94, 1420 (1954); 97, 641 (1955).
5. Collection: Physics of Semiconductor Surfaces, ed. by Pikus.[Russian translation] (IL, 1959).
6. J. Bardeen, R. E. Coovert, et al., Phys. Rev. 104, 47 (1956).
7. A. Many, E. Harnik, and Y. Margoninski, Semiconductor Surface Physics (New York, 1957), p. 85.

EFFECT OF SOME CHEMICAL TREATMENTS ON THE SURFACE PROPERTIES OF GERMANIUM AND ON THE PARAMETERS OF SEMICONDUCTING DEVICES

V. F. Synorov, V. V. D'yakov, and L. I. Bobrova

V. V. Kuibyshev State University, Tomsk

An interesting possibility in the production of stable semiconducting devices is the so-called "passivation" of their surfaces [1]. The problem is to produce a thin protective layer on the semiconductor surface, which:

1) should be impermeable to moisture, oxygen and other constituents of the atmosphere;

2) should not impair the parameters of the semiconducting device;

3) should be dielectric or near-dielectric in its properties;

4) should be stable in time and be unaffected by temperature variations;

5) should be mechanically strong, etc.

Experience has shown [2-4] that the instability and deterioration of the parameters of a germanium transistor are, to a large extent, due to the instability of the surface oxide layer, to its permeability (which is related to its amorphous and loose structure), and to its considerable solubility. This leads us to the problem of a deliberate chemical change of the state of a germanium surface during the manufacture of devices, i.e., the formation on the germanium surface of stable chemical compounds which include germanium as one of the components and which are continuations of the germanium lattice.

Among the various possible ways of forming stable surface compounds on germanium for the purpose of isolating it from the ambient medium, the formation of germanium sulfides is of special interest. It is known [5] that germanium sulfides obtained in the form of a separate phase are quite stable under the action of oxygen and water at temperatures up to 300°C and above.

The process of sulfurization of some metals in order to give their surface greater mechanical strength and corrosion resistance is widely used in practice [6].

For the sulfurization of a germanium surface, we used a special bath with the following composition:

a) main component - chemically neutral salts with low melting points;

b) active component - compounds of sulfuric acid in which the sulfur atoms possess reducing properties;

c) catalyst salt for the continuous support of the reducing properties of the active salt.

This eutectic bath had a melting point of about 420°C. Sulfurization of germanium samples (which were first etched, degreased and washed) was carried out at 430-450°C for 20-30 min. Then the samples were washed several times in hot, doubly-distilled water and dried.

The sulfurized germanium surface was lightly colored and the sulfide layer was very dense and continuous. Figure 1 shows photomicrographs of the surface after sulfurization and after treatments in hydrochloric and hydrofluoric acids. It is clear that these two acids have no marked effect on the sulfide layer.

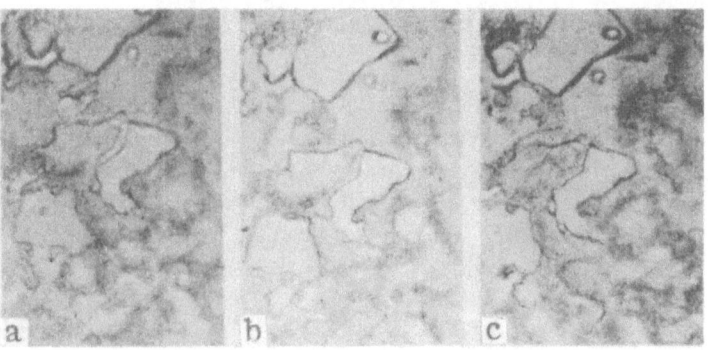

Fig. 1. Photomicrographs of the (111) surface of germanium; magnification of 450. a) After sulfurization; b) after treatment of the surface a in hydrochloric acid; c) after treatment of the surface a in hydrofluoric acid.

TABLE 1

Samples	Microhardness, kg/mm^2			Mean standard deviation	
	Average	Maximum	Minimum	Absolute value	%
Control	229	311	175	49	20.6
Sulfurized	463	650	358	98	21.3
Sulfurized and vacuum-heated	430	487	385	40	9.3
Sulfurized and heated in nitrogen	501	578	417	61	12.3

We also investigated some physico-chemical and electrical properties of the sulfurized germanium surface. Moreover, we heated some sulfurized samples in vacuum in order to find what changes in the structure and properties of the layer occur when it fuses into germanium, and some further samples were heated in an atmosphere of nitrogen at 450°C. Table 1 lists the results of microhardness tests on sulfurized samples using the metallographic instrument PMT. Vacuum heating made the mechanical properties of the surface more uniform and nitration reduced the scatter and raised the microhardness values. The "control" samples referred to in Tables 1 and 2 represent a batch of the usual germanium samples with etched surfaces.

The study was carried out on n-type germanium of $2 \Omega \cdot$ cm resistivity. Batches of 30-40 samples were used. The microhardness was measured at different points on the surface. Table 1 lists the average maximum and minimum values and the quantities which represent the scatter of these values.

To find the change in the degree of hydration of the surfaces, we determined the relative gain in weight of the samples after storage in an atmosphere of 100% relative humidity for 24 hours. For this, we prepared

160

TABLE 2

Sample	Relative gain in weight after 24-hour storage in humid atmosphere, %
Control	0.128
Sulfurized	0.0098
Sulfurized and vacuum-heated	0.001
Sulfurized and heated in nitrogen	0.015

special batches of control and sulfurized samples so that the total surface of the samples in a batch should be not less than 30-40 cm². The samples were subjected to preliminary drying in a dry atmosphere at 85-90°C. Weighing was carried out with a analytical microbalance having an optical indicator. The relative gain in weight per unit area of the surface is given in Table 2.

The sulfide layer thickness was determined optically, and, depending on the duration of sulfurization, it varied from 2 to 4 μ.

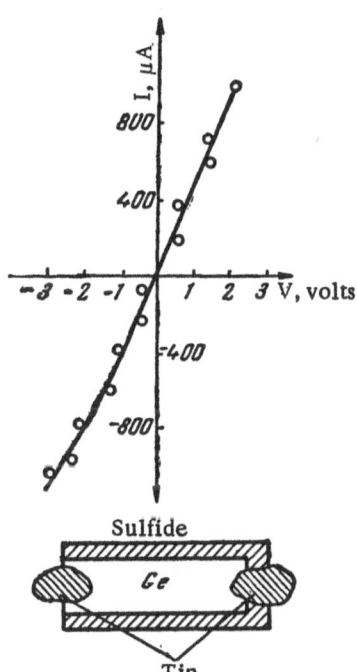

Fig. 2. Structure of a sample of sulfurized germanium with tin contacts and its current voltage characteristic.

The stability of the sulfurized germanium surface when the ambient medium was changed was checked by measuring the contact potential difference with respect to a platinum reference electrode, using the capacitor method. In contrast to the etched but not sulfurized germanium surface, the contact potential difference between the sulfurized surface and platinum was constant, within the experimental error (± 0.01 V), during prolonged storage at room temperature in air with varying humidity and also when the pressure was lowered to 5 × 10^{-6} mm Hg.

According to the results obtained by the photomagnetic method, the surface recombination velocity of the sulfurized germanium was 44-64 cm/sec and varied within 10-15% after exposure to water vapor. Measurements of the field effect showed that the sulfurized germanium surface had a negative charge of (16.1-9.5) × 10^{-10} C/cm² density. These results are in accord with the data on the surface recombination velocity and, according to the theory of Stevenson and Keys [7], account for the low value of this velocity.

The results obtained suggest that the process of sulfurization can be used to "passivate" the surface of germanium in the initial stage of manufacture of germanium devices.

To check that the contacts, prepared by fusing in tin through the sulfide layer, are ohmic we used special samples, as shown in the lower part of Fig. 2. On one side of the sample, the tin was fused through the sulfide layer and on the other the layer was ground off and the usual ohmic contact established.

Figure 2 shows the current-voltage characteristic of such a sample. These preliminary experiments demonstrate that an ohmic contact with sulfurized germanium can be obtained simply by fusing in tin without additional technological processes.

To find, in principle, whether it is possible to manufacture transistors from sulfurized germanium, we carried out tests on p-n-p transistors made by fusing in indium in vacuum at 550°C. The indium was fused in

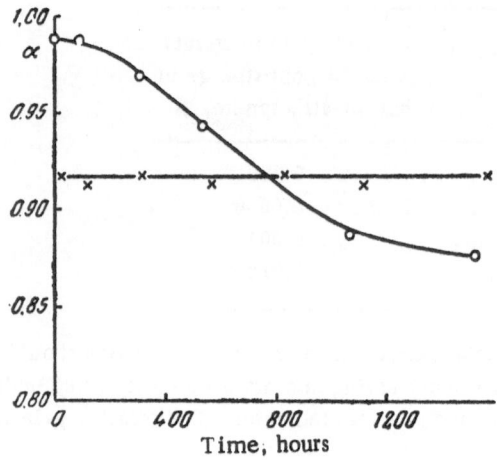

Fig. 3. Dependence of the current amplification factor on the duration of storage at room temperature. Each point represents an average value for 12 transistors: O - control samples; × - sulfurized samples.

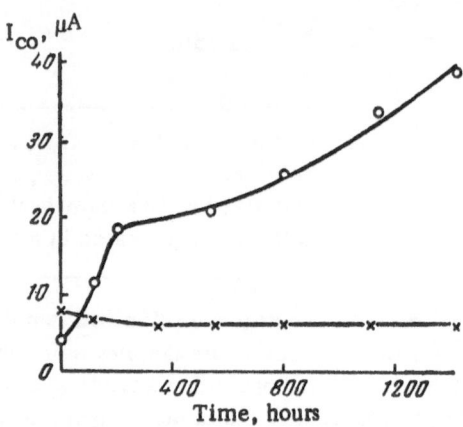

Fig. 4. Dependence of the reverse collector current on the duration of storage at room temperature. Each point represents an average value for 12 transistors: O - control samples; × - sulfurized samples.

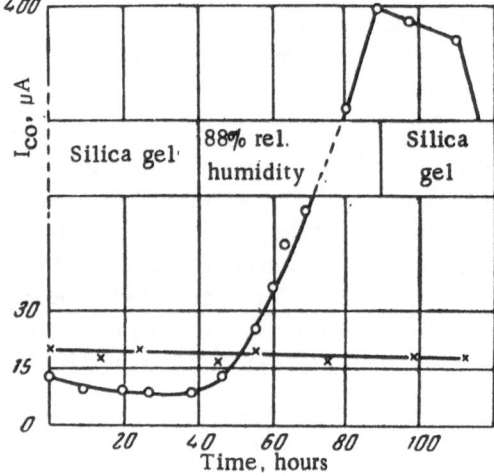

Fig. 5. Variation of the reverse collector current with relative humidity. Each point represents an average value for 6 transistors: O - control samples; × - sulfurized samples.

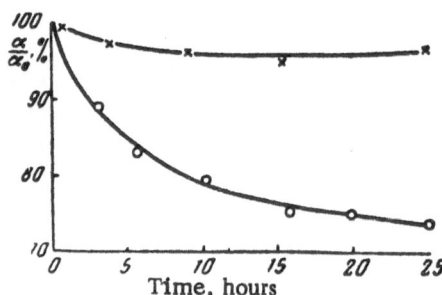

Fig. 6. Relative change of the current amplification factor after heat treatment at 65°C. Each point represents an average value for 10 transistors; O - control samples; × - sulfurized samples.

through the sulfide layer without any additional treatment, except for brief etching of the samples in Perhydrol in order to remove surface contamination. Several batches of samples were prepared.

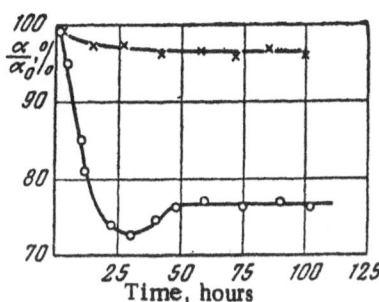

Fig. 7. Relative change of the current amplification factor after heat treatment at 85°C. Each point represents an average value for 10 transistors: ○ - control samples; × - sulfurized samples.

To discover the influence of the sulfurized surface on the parameters of germanium p-n-p transistors, we measured the current amplification factor α and the reverse collector current I_{co} under various conditions of sample storage. For the sake of comparison, we also measured the parameters of unprotected transistors prepared in the same way (except for sulfurization). The results of the measurements are shown in Figs. 3-7.

Figures 3 and 4 show the average values of the transistor parameters as a function of the duration of storage under room conditions. It is seen that the sulfurized surface has protective properties.

Figure 5 records the dependence of the average reverse collector current of the control and the sulfurized samples when the relative humidity was varied. Initially, the samples were kept in a closed vessel with dry air (dried by means of silica gel), then they were placed in a humid chamber and again in a dry chamber. The results show that the sulfurized samples withstood well the influence of water vapor at room temperature.

It is known that the process of "aging" of transistors and the reversible drift of the amplification factor are related to surface properties: the continuing process of oxidation and the variation of the degree of hydration of the surface. In practice, artificial aging is carried out and the devices are checked by the so-called 48-hour effect (reversible drift).

Also we carried out heat treatment of the control and sulfurized triodes at 65 and 85°C. Figures 6 and 7 give the relative changes, α/α_0, where α is the current amplification factor before heat treatment. The parameters were measured 15 min after the removal of the device from a thermostat, i.e., they were measured at room temperature. From the results obtained we may conclude that the sulfide layer on the surface of germanium protects this surface from the continuing processes of oxidation and from the hydration—dehydration processes which takes place on changing the temperature from room to higher values and back again to room temperature.

In conclusion, we note that the parameters of the transistors made from sulfurized germanium were somewhat poorer than those of the control samples. This may be due to the insufficiently developed technique of fusing in indium to the sulfide layer and due to some nonuniformity of the surface which lowers the average values of the parameters.

The preliminary experiments lead to the following conclusions.

1. In principle, it is possible to form a stable chemical compound on the surface of germanium by sulfurization.

2. A sulfurized germanium surface has better mechanical properties and is less affected by hydration than a free surface.

3. The electrical properties of sulfurized surfaces are quite stable and compare satisfactorily with those obtained after other treatments, for example, etching.

4. Ohmic contacts can be obtained by fusing in tin through the sulfide layer.

5. In principle, we can prepare p—n junctions by fusing in indium through the sulfide layer.

6. The parameters of the test transistors are stable under the action of the atmosphere and water vapor at room and elevated temperatures.

LITERATURE CITED

1. Zarubezhnaya radioélektronika No. 10, 136 (1960).
2. J. Wallmark, RCA Rev. $\underline{18}$, 243 (1957).
3. J. Wallmark and P. Johnson, RCA Rev. $\underline{18}$, 512 (1957).
4. R. Kingston, J. Appl. Phys. $\underline{27}$, 101 (1956).
5. O. Johnson, Chem. Rev. $\underline{51}$, 431 (1952).
6. M. Neely, La Machine Moderne 47 (1953); 41 (1953).
7. A. Stevenson and R. Keys, Physica $\underline{20}$, 1041 (1954).

EFFECT OF THE SURFACE STATE IN SILICON
p-n JUNCTIONS ON THE REVERSE CURRENTS
AND DRIFT

R. O. Litvinov and Hsü Tung-liang

Institute for Semiconductors, Academy of Sciences, UkrSSR

A new way of treating the surface of silicon p-n junctions was used to determine the nature of excess reverse currents and the appearance of drift on the application of a bias.

It is usually assumed that the rise of the reverse currents and their instability are due to the action of adsorbed moisture. However, the problem of the mechanism of the influence of water is still largely unsolved [1]. The difficulty lies in the ambiguous interpretations of the observed phenomena. Therefore, the use of such techniques as the field effect in samples with p-n junctions and adsorption of gases, which, like water, produce positive surface charge in addition to the intrinsic charge, should help to solve this problem. Each of these methods has certain features in common with the effect of water on the semiconductor properties but each also has other different features.

The reverse current of pulled p-n+ junctions and the conductivity of the high-resistivity p-type parts of the samples were investigated under the influence of a constant external electric field in vacuum. This allowed us to determine the surface potential and its changes under the action of the external field, as well as the changes of the surface recombination velocity. Comparison of these results with measurements of the reverse current and the photo-emf showed that channel leakage occurred under the action of positive fields. The large changes in the reverse current could not be explained by changes in the surface recombination velocity [2]. The magnitude of the excess current indicated the necessity to allow for the current of the channel space-charge [3]. It should be noted that in vacuum in the presence of a channel no drift of the reverse current was observed.

We investigated also the effect of adsorption of ammonia and water vapor. The adsorption of ammonia increased the reverse current and produced drift. The surface potential changed in the direction of formation of the channel. There was no correlation between the thickness of the oxide and the time constants of the current drift, which should have been obtained if the drift were due to the establishment of equilibrium across the oxide layer between the surface ("slow") states and the interior of the semiconductor after the application of a bias [4].

In water vapor (~75% rel. humidity) the current increased even more and the drift also rose sharply. After the bias was removed, the equivalent capacitance and the photo-emf of p-n junctions decreased slowly, approaching the initial values.

From the analysis of our results, we concluded that the positive surface charge appearing on adsorption of water vapor and ammonia produces a surface channel in the high-resistivity p-type parts of the samples, in the same way as the action of a positive external field in vacuum.

The current drift is apparently related to the modulation of the channel by slow variation of the surface charge after application of the bias. Most probably this charge changes due to direct displacement of the adsorbed ions along the surface.

Thus the excess currents and the drift in silicon devices are mainly due to the appearance and variation of channel leakage.

In the case of saturated water vapor, a very strong (by two orders of magnitude) increase of the current was observed, as well as a change in the nature of the drift (decay of the current with time) and poor reproducibility of the results. This is due to the increased role played by leakage in the adsorbed water layer.

A detailed report of these results was published elsewhere [5].

LITERATURE CITED

1. R. Solomon, J. Appl. Phys. 31, 1791 (1960).
2. J. H. Forster and H. S. Veloric, J. Appl. Phys. 30, 906 (1959).
3. C. T. Sah, R. N. Noyce, and W. Shockley, Proc. Inst. Radio Engrs. 45, 1228 (1957).
4. M. Lasser, C. Wysocki, and B. Bernstein, Phys. Rev. 105, 491 (1957).
5. R. O. Litvinov and Hsü Tung-liang, Radiotekh. i Élektron, 7, 1030 (1962).

INDEX